NANOSTRUCTURED SEMICONDUCTORS IN POROUS ALUMINA MATRICES

Modeling, Synthesis, and Properties

NANOSTRUCTURED SEMICONDUCTORS IN POROUS ALUMINA MATRICES

Modeling, Synthesis, and Properties

Rishat G. Valeev, PhD

Alexander V. Vakhrushev, DSc

Aleksey Yu. Fedotov, PhD

Dmitrii I. Petukhov, PhD

Editorial Board Members

A. N. Beltiukov, A. L. Trigub, and A. V. Severyukhin

Apple Academic Press Inc.
3333 Mistwell Crescent
Oakville, ON L6L 0A2
Canada USA

Apple Academic Press Inc.
1265 Goldenrod Circle NE
Palm Bay, Florida 32905
USA

Exclusive worldwide distribution by CRC Press, a member of Taylor & Francis Group

International Standard Book Number-13: 978-1-77188-770-0 (Hardcover)
International Standard Book Number-13: 978-0-42939-814-8 (eBook)

Library and Archives Canada Cataloguing in Publication

Title: Nanostructured semiconductors in porous alumina matrices : modeling, synthesis, and properties / Rishat G. Valeev, PhD, Alexander V. Vakhrushev, DSc, Aleksey Yu. Fedotov, PhD, Dmitrii I. Petukhov, PhD ; editorial board members, A.N. Beltiukov, A.L. Trigub, and A.V. Severyukhin.

Names: Valeev, Rishat G., author. | Vakhrushev, Alexander V., author. | Fedotov, Aleksey Yu., author. | Petukhov, Dmitrii I., author.

Description: Includes bibliographical references and index.

Identifiers: Canadiana (print) 20190113022 | Canadiana (ebook) 20190113227 | ISBN 9781771887700 (hardcover) | ISBN 9780429398148 (PDF)

Subjects: LCSH: Nanostructured materials. | LCSH: Semiconductors. | LCSH: Aluminum oxide. | LCSH: Aluminum films.

Classification: LCC TA418.9.N35 V35 2019 | DDC 621.3815/2—dc23

CIP data on file with US Library of Congress

Apple Academic Press also publishes its books in a variety of electronic formats. Some content that appears in print may not be available in electronic format. For information about Apple Academic Press products, visit our website at **www.appleacademicpress.com** and the CRC Press website at **www.crcpress.com**

About the Authors

Rishat G. Valeev, PhD, is a Senior Researcher in the Department of Physics and Chemistry of Surface of the Physical-Technical Institute of the Ural Branch of the Russian Academy of Sciences. Dr. Valeev has over 200 publications to his name, including chapters in monographs, articles, reports, reviews, and patents. He managed several scientific projects, including one funded by the Russian President, projects of the Russian Federal Programs, and projects of the Program of Presidium of the Russian Academy of Sciences. His research interests include light-emitting semiconductor nanostructures, EXAFS-spectroscopy, development of physical vapor deposition methods for evaporation of semiconductors and metals, application of porous alumina for synthesis of nanostructures, cathodic hydrogen production, and investigation of basic laws relating to the structure and macro characteristics of nanostructures. He is an expert of the Russian Science Foundation and is affiliated with the *Journal of Structural Chemistry, Semiconductors Science and Technology, Materials Research Express*, and others.

Alexander V. Vakhrushev, DSc, is the Head of the Department of Mechanics of Nanostructures of the Institute of Mechanics of the Ural Branch of the Russian Academy of Sciences and Head of the Department of Nanotechnology and Microsystems of Kalashnikov Izhevsk State Technical University. Dr. Vakhrushev is a corresponding member of the Russian Engineering Academy. He has over 400 publications to his name, including monographs, articles, reports, reviews, and patents. He has received several awards, including an Academician A. F. Sidorov Prize from the Ural Division of the Russian Academy of Sciences for significant contribution to the creation of the theoretical fundamentals of physical processes taking place in multi-level nanosystems and Honorable Scientist of the Udmurt Republic. He is currently a member of editorial board of several journals, including *Computational Continuum Mechanics, Chemical Physics and Mesoscopia, and Nanobuild*. His research interests include multiscale mathematical modeling of physical-chemical processes into the nano-hetero systems at nano-, micro-, and macro-levels; static and dynamic interaction of nanoelements; and basic laws relating the structure and macro characteristics of nano-hetero structures. He has published several books, including *Computational Multiscale Modeling of Multi-phase Nanosystems: Theory and Applications, Theoretical Foundations and Application of Photonic Crystals*, and *Nanomechanics*.

Aleksey Yu. Fedotov, PhD, is a Senior Researcher in the Department of Mechanics of Nanostructures of the Institute of Mechanics of the Ural Branch of the Russian Academy of Sciences and docent of the Department of Nanotechnology and Microsystems of Kalashnikov Izhevsk State Technical University. Dr. Fedotov's main area of scientific interest is the study of the processes of formation and interaction of nanoparticles and the study of the dependence of the properties of composite materials on the quantitative, dimensional, and structural parameters of inclusions. He has over 80 publications to his name, including monographs, articles, reports, reviews, and patents. He has received several awards, including an Academician A. F. Sidorov Prize from the Ural Division of the Russian Academy of Sciences for the best work in applied mathematics on the topic "Creation of a software complex for multilevel mathematical modeling of physical and chemical processes in nanomaterials."

Dmitrii I. Petukhov, PhD, is a Researcher in the Department of Physics and Chemistry of Surface of the Physical-Technical Institute of the Ural Branch of the Russian Academy of Sciences. Dr. Beltiukov has about 100 publications to his name. His research interests include synthesis of semi-conductor nanostructures and nanocomposite systems such as nanowires, quantum dots, and nanocrystallites in an oxide matrix, investigation of their structure, morphology, optical characteristics, and electron structure.

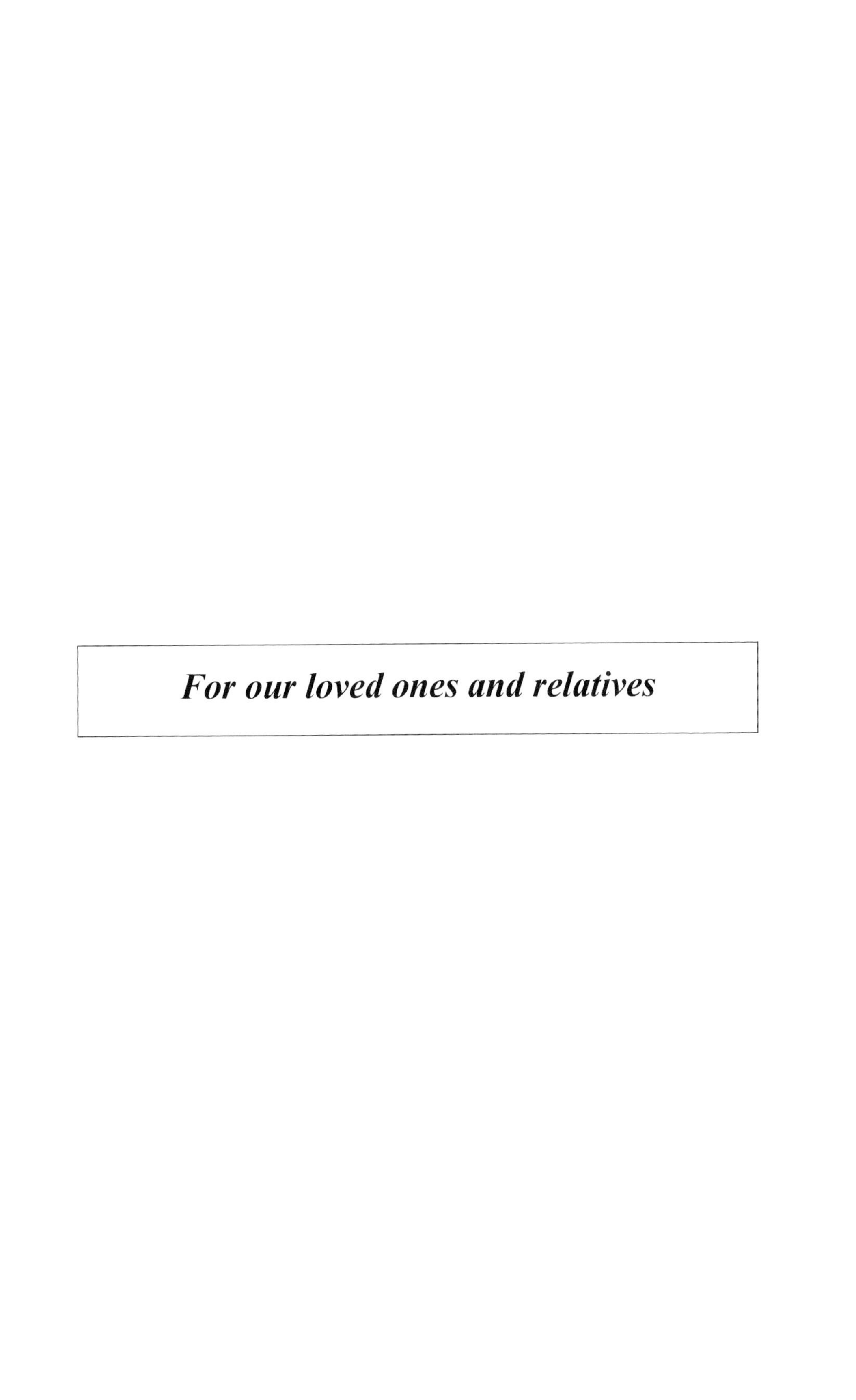

For our loved ones and relatives

Contents

Abbreviations ..... *xi*

Preface ..... *xiii*

Introduction ..... *xv*

Acknowledgments ..... *xix*

Epilogue ..... *xxi*

1. **Brief Overview of the Literature on Nanostructural Electroluminescent Light Sources 1**
2. **Porous Anodic Aluminum Oxide: Structure, Properties, and Application in Semiconductor Technology 19**
3. **Vacuum-Thermal Deposition of Semiconductors 37**
4. **Methods of Investigating Structure and Chemical Composition 51**
5. **Methods of Optical Spectroscopy 73**
6. **Theoretical Models for Investigating The Processes of Nanofilm Deposition onto Porous Templates of Aluminum Oxide 85**
7. **Synthesis of Electroluminescent Nanostructures of ZnS Doped by Cu and Mn Ions 141**
8. **Structure and Chemical Composition of Electroluminescent Nanocomposites ZnS:(Cu,Mn)@AAO 165**
9. **Optical Properties of Electroluminescent Nanostructures of Cu and Mn-Doped ZnS 191**
10. **Results of Modeling the Deposition Processes of Nanofilms onto Aluminum Oxide Templates 205**

Index* *247

Abbreviations

AAO	anodic aluminum oxide
AC	alternating current
AMBER	Assisted Model Building with Energy Refinement
CSR	coherent scattering regions
CVD	chemical vapor deposition
CVFF	consistent valence force field
DC	direct current
DFT	density functional theory
DPs	dispersed particles
EAM	embedded-atom method
ELE	electroluminescent emitter
ELS	electroluminescent light sources
EXAFS	extended X-ray absorption fine structure
FWHM	full width at half maximums
GROMOS	GROningen MOlecular simulation
ITO	indium tin oxide
JCPDS	Joint Committee on Powder Diffraction Standards
LB	Langmuir–Blodgett
MBE	molecular beam epitaxy
MD	molecular dynamics
MEAM	modified embedded-atom method
MOCVD	metal-organic chemical vapor deposition
MS	multiple scattering
NEXAFS	near-edge X-ray absorption fine structure
NOP	normalized oscillation part
OR	Oswald ripening
PET	polyethylene terephthalate
PL	photoluminescence
QCM	quartz crystal monitor
QM	quantum mechanics
REE	rare-earth elements
REM	raster electron microscopy
RSF	relative sensitivity factor

SEM	scanning electron microscopy
TEM	transmission electron microscopy
TFELS	thin-film electroluminescent sources
UHV	ultrahigh vacuum
UV	ultraviolet
XAFS	X-ray absorption fine structure
XANES	X-ray absorption near-edge structure
XPS	X-ray photoelectron spectroscopy

Preface

This book deals with a systematic description of the complex experimental and theoretical investigation of functional semiconductor nanostructures in porous anodic alumina matrices. Much consideration is given to the processes of the formation of functional semiconductor nanostructures in porous anodic alumina matrices. The book concentrates on the description of experimental methods, the results of experiments on the formation of nanostructures in porous anodic alumina matrices, and the study of their various physical properties. The book includes, also, several mathematical models of nanofilm formation on porous templates of aluminum oxide and investigates the influence of dimensional parameters of the pores in aluminum oxide matrix on the processes of formation of nanofilm coatings by the molecular dynamics method. Gold, silver, chromium, copper, iron, gallium, germanium, titanium, platinum, vanadium, and other atoms were used as precipitated materials in the experiments and while modeling.

The information presented can be useful to engineers, researchers, and postgraduate students engaged in the design and research in the field of nanotechnology.

Introduction

At present, studies devoted to the synthesis of various nanostructured materials are very relevant. Promising classes of such materials are functional semiconductor nanostructures based on 0D (quantum dots and free and stabilized nanoclusters), 1D (nanothreads, nanorods, nanotubes, nanobands, and whiskers), 2D (fine films up to several hundred nanometers, heterostructures, and Langmuir–Blodgett films), and 3D (nanoparticles, including those in envelopes, and 3D self-organizing arrays of nano-objects) are actively studied. Luminescent nanoparticles embedded into and inorganic dielectric matrixes are of special interest, since the matrix allows not only spatially dividing separate nanoparticles but also protecting them from the environment. The films of porous alumina were the most widely used as matrices for the synthesis of ordered arrays of nanostructures of various compositions. Light-emitting electroluminescent panels based on powders and fine zinc sulfide electroluminophors are widely applied, mainly, in devices of symbol information display, illumination panels of advertising structures. Modern methods of applying materials, such as electrochemical deposition, luminescent ink printing, chemical vapor deposition, various options of vacuum-thermal precipitation allow forming electroluminescent layers practically on any surface.

Therefore, experimental and theoretical studies of the processes of formation of these nanostructures and their properties are very relevant.

This book comprises 10 chapters.

Chapter 1 contains the brief overview of the literature on nanostructural electroluminescent light sources. The descriptions of electroluminescent light sources (ELS), electroluminescence mechanism in zinc sulfide ELS and luminophores based on nanostructures of alloyed zinc sulfide are given.

Chapter 2 is devoted to porous anode aluminum oxide and its structure, properties, and application in semiconductor technology. Types of aluminum oxide films formed during anodizing, structure of anode aluminum oxide films, mechanism of porous structure formation, and influence of anodizing parameters on the structure of anode aluminum oxide film are described in detail in this chapter. Also, the use of porous aluminum oxide to obtain nanostructures of semiconductors is presented at the end of the chapter.

Chapter 3 contains the description of the methods of the vacuum-thermal deposition of semiconductors. Processes of evaporation and deposition are described. Experimental approaches and equipment for vacuum-thermal deposition and evaporators are also described in detail in this chapter.

Chapter 4 describes methods of investigating structure and chemical composition. In this chapter, X-ray diffraction, EXAFS-spectroscopy, and application of Fourier fitting to extract the structural information from EXAFS spectra and X-ray electron spectroscopy are discussed.

Chapter 5 is dedicated to the detailed description of the methods of optical spectroscopy which we used in optical investigations. Monochromator on diffraction grids and operating principle on diffraction grids are considered. The technique of determination of the band gap width of semiconductors by the spectra of optical transmission is presented. The main illuminating characteristics of electroluminescent emitters and devices and methods for measuring electroluminescent emitter radiation characteristics are also described in this chapter.

Chapter 6 contains mathematical models of nanofilm formation on porous templates of aluminum oxide. The modeling methods of the problem investigated are studied. The equations describing the method of molecular dynamics, as well as the equations of nanoparticle movement in mesic media are demonstrated. The periodic boundary conditions and their influence on the system modeled are formulated. The potentials of different interaction types emerging in the system modeled are described. Different numerical algorithms for solving the modeling problem of nanofilm coating formation are given. The formulas for calculating thermodynamic parameters of the system modeled are analyzed.

Chapter 7 presents the results of experimental investigations of synthesis of electroluminescent nanostructures of ZnS alloyed by Cu and Mn ions. At first, the synthesis of AAO porous matrixes is described. The results of coulometric control of oxide film thickness and pore diameter control are given. Then the results of alloyed zinc sulfide deposition are presented. In this part of chapter, the facility of ultrahigh-vacuum evaporation based on the vacuum system USU-4 are described and results of experimental investigations of deposition of ZnS alloyed with Cu and Mn onto smooth templates and matrixes of porous AOA are presented and discussed.

Chapter 8 includes the results of complex experimental investigations of the structure and chemical composition of electroluminescent nanocomposites ZnS:(Cu,Mn)@AAO. Chemical composition and chemical bond character of such nanocomposites are discussed. The results of XPS experiments of samples alloyed with Cu and Mn are presented. Morphology of nanostructures based on raster electron microscopy data, X-ray diffraction investigations, and local atomic structure based on EXAFS spectroscopy data is considered. In addition, the results of EXAFS investigations of samples alloyed with Cu and Mn and results of XANES investigations are described.

In *Chapter* 9, the experimental results of optical properties of electroluminescent nanostructures of ZnS alloyed with Cu and Mn ions are presented. Absorption spectra and optical band gap and electroluminescence are also studied in this chapter.

Chapter 10 presents the simulation results of modeling the deposition processes of nanofilms onto aluminum oxide templates. At first, setting up and stages of solving the problem of nanofilm deposition are described. Then, the simulation results of formation of monofilms on porous alumina substrates are presented. Also, the effects of pore size into substrates on the nature of the deposited nanofilm are studied. Investigation of the structure of the templates and nanofilms, and formation of multicomponent nanofilms and films with admixtures on the templates of porous aluminum oxide are presented. In the end of chapter, the control of the processes of atom deposition and formation of nanofilm coatings are discussed.

The epilogue focuses on three main issues: the main results of simulation, experimental investigations, as well as further perspectives for future research.

Acknowledgments

The authors express the sincerest gratitude to Academician A. M. Lipanov for his fruitful cooperation, counseling, constant support, and attention to the work.

The authors are grateful to their colleague Severyukhina O. Yu., PhD, for his active participation in the development of the software complex, participation in numerous calculations, and analyzing of simulation results. We would also like to express gratitude A. I. Chukavin and V. M. Vetoshkin for their active participation in experiments and analysis of experimental results.

The main research results (the mathematical models and computational and experimental investigations) within the framework of the project No. 15-19-10002 of Russian Science Foundation were obtained.

The investigation was carried with financial support from the Ural Branch of the Russian Academy of Sciences (projects No. 15-10-1-23 and No. 18-10-1-29), the programs of the Russian Academy of Sciences Presidium and the Kalashnikov Izhevsk State Technical University.

Within the framework of the Ural Branch of the Russian Academy of Sciences projects, the simulation of the formation of porous nanomaterials was carried out.

The calculations were performed at the Joint Supercomputer Center of the Russian Academy of Sciences.

Epilogue

The epilogue focuses on three main issues: the main results of simulation, experimental studies, as well as prospects for future research.

SIMULATION

Molecular-dynamic methods of modeling precipitation processes allowed illustrating the variants of epitaxial overgrowing of porous substrates based on alumina by various types of atoms, various processes of interaction of nanostructures and mechanisms of substrate and pore growth depending on the parameters of the porous structure (pore diameter and distances between them), angles tilt of the molecular beam relative to the substrate, as well as subsequent annealing at various temperatures.

For different types of deposited atoms, various processes of interaction of nanostructures and mechanisms for the inoculation of substrates and pores with a diameter of 2 nm were recorded. In general, a uniform coating of the pores with a nanofilm with a partial penetration of the atoms inside was observed. The maximum filling was observed when the atoms of gallium and germanium were overgrown.

When studying the inversion of gallium atoms with coatings with pores of various sizes, it was found that the active growth of the number of atoms in the pore occurs in the initial period of time. Stabilization of the center of mass of the deposited atoms occurs at different depths of the pore. For pores of radius 2 and 3 nm, the center of mass is formed above the middle of the depth of the pore. With increasing pore size, the center of mass begins to form in one place, near the middle of the depth of the pore.

In the precipitation of zinc sulfide, the analysis of the structure of the materials was carried out on the basis of the average value of the ideality parameter of the crystal lattice. The structure of both the substrate and the formed nanofilm was evaluated. It is obtained that the parameter value for both substrates and nanofilms is sufficiently high, which indicates an amorphous structure of the formed materials. The end of the deposition of nanofilms is key to changing the structure: if before it, the restructuring of the structure and the motion of the atoms took place actively, then after it, the change of coordinates to a more energetically favorable position is insignificant. In the process of condensation, the smallest value of the crystal structure parameter has a computational experiment without pores with the addition of 5% copper atoms. The stabilized sputtered material has the smallest parameter value for the case of a substrate without pores and without the addition of copper atoms.

In order to obtain results that are as close as possible to the experimental data, the dimensions of the modeling area have been significantly increased. The pores in the alumina substrate was cut by a radius of 5 nm and a depth of 10 nm. The proportion of alloying elements increased the proportion of epitaxial atoms proportionally to their composition. In general, the process of overgrowing can be formally divided into two stages: partial filling of the cavity in the substrate with a uniform growth of nano-formations, and sometimes forming a "cover" covering the entry of new deposited atoms into the pores.

Simulation of deposition processes at different angles of incidence of a molecular beam with respect to the normal direction of epitaxy has shown that the deposition process is insignificantly dependent on the angle and is practically identical in time. The deviation from the normal to the surface of the substrate leads to an insignificant increase in the area to which the atoms are directly deposited.

To investigate the mechanisms of sintering nanofilms on porous alumina substrates, already sprayed coatings of molecular zinc sulfide from the previous series of computational experiments were considered. The total duration of the heating-up period of the substrate-nanofilm nanosystem was 200 ps, simulations were carried out for temperatures of 293 K, 493 K, and 593 K. In all the cases, atoms, nanofilms, and other composite nanostructured objects behave identically in the system.

EXPERIMENTS

A procedure for the synthesis of porous matrices of anodic alumina with specified thickness and parameters of the porous structure (pore diameter and distance between them) has been developed.

Since the process of formation of the oxide film is electrochemical, the mass of the formed oxide film, and, consequently, its thickness is determined by the charge that has leaked during the anodizing process, so a method of coulometric control was used for precise control of the thickness of the oxide film.

To determine the coefficient that connects the charge density passed through the anodizing process to the thickness of the oxide film, films of anodic aluminum oxide of various thicknesses were synthesized by passing a fixed charge at voltages of 40, 80, and 120 V. The thickness of the oxide film is proportional to the charge passed through the anodizing process, and the proportionality coefficient increases with the anodization voltage from 0.415 ± 0.005 (for a voltage of 40 V) to 0.550 ± 0.030 ($\mu m \cdot cm^2$)/C (for a voltage of 120 V). The change in the proportionality coefficient, which connects the missed charge and the thickness of the oxide film, can both with different rates of side reactions reduces the current efficiency with decreasing anodization voltage. In addition, a possible reason for the variation in the coefficient may be a change in the porosity of the film or a change in the density of aluminum oxide formed at different voltages.

The pore diameter of the anodic alumina can be varied by varying the anodizing voltage; however, with increasing anodization voltage, the porosity of the oxide film is maintained at approximately 10%. In order to increase the porosity of the oxide matrix and, consequently, to increase the degree of filling of the oxide matrix to increase the pore diameter, it has been proposed to use the method of chemical etching in a solution of phosphoric acid. Based on the analysis of the dependence of the average pore diameter on the duration of etching, it is established that the rate of pore rupture is 0.54 nm/min. Etching occurs to some value of the average pore diameter; further, etching leads to the destruction of the honeycomb frame, which consists of a denser alumina and complete destruction of the porous structure.

On the basis of the USU-4 ultrahigh vacuum unit, a setup of an ultrahigh vacuum (no worse than 10^{-7} Pa) deposition equipped with a chamber for substrates loading with an independent pumping system, an analytical chamber, and a precipitation chamber for zinc-containing compounds equipped with three Knudsen evaporation cells with crucibles from pyrolytic boron nitride with indirect heating of crucibles, flaps and thermocouples for controlling the temperature of the crucible is done. This made it possible to realize the connection between the power sources of the evaporative cell heaters and the thermocouples to accurately maintain the set temperature with the help of proportional–integral–derivative (PID) regulators and set the operating modes of the evaporators programmatically. The amount of precipitated material is monitored by means of a quartz thickness gauge.

A technique for the thermal deposition of doped zinc sulfide on porous alumina substrates has been developed with the aim of synthesizing luminescent nanostructures. Doping was carried out by spraying zinc sulfide and a dopant from various molecular sources. The concentration of the alloying element (copper and manganese, and also copper and manganese with coactivator in the form of chlorine atoms) was set by the temperature of the crucible with the material of the alloying element. For copper alloying, copper was used. For manganese doping, manganese sulfide was used. Chlorides of copper and manganese were used for doping with the coactivator.

X-ray diffraction, X-ray photoelectron spectroscopy (XPS), and extended X-ray absorption fine structure spectroscopy have shown that when doping with copper with a concentration of no more than 5 at.%, the Cu_2S phase is formed, the presence of which is one of the causes of electroluminescence. The resulting ZnS: Cu nanostructures crystallize in the cubic phase with a small admixture of the hexagonal phase. Thus, upon deposition on a porous alumina of 1 μm thickness, a separate phase of copper is formed in the form of nanoclusters. When manganese is doped with manganese sulfide phases, it is not isolated but is embedded in the lattice instead of zinc atoms. The change in the concentration of the alloying element leads to changes in the lattice parameter and, as a result, to the appearance of distortions at the local level.

Doping with copper together with chlorine leads to the separation of the copper phase, presumably at the semiconductor/matrix interface, since, according to the XPS data, there is no pure copper on the surface, and copper spikes are observed. The chlorine has no other effect on the structure of the obtained samples. In the case of doping with manganese and a coactivator, the presence of chlorine does not affect the structure of the obtained samples. From this, it can be concluded that the chlorine atoms either replace the sulfur atoms in the lattice, or, more likely, are in the interstitial space.

Prototypes of light-emitting devices are made and their properties are investigated. For samples without the ZnS: Cu and ZnS: Mn coactivator (5 and 10 at.%) deposited on porous alumina matrixes with 40, 80, and 120 pore diameters, the maximum radiation is observed for samples obtained on matrices with a pore diameter of 80 nm. Magnifying the concentration of the alloying element leads to "quenching" of the luminescence, which is associated with an increase in the number of crystal lattice defects in the crystal phosphorus and is well known.

In the electroluminescent light sources (ELS) ZnS: Cu spectra without a coactivator, maxima with wavelengths of 526 and 556 nm can be distinguished, and in ZnS: Mn—449, 520, and 566 nm. The maximum with a wavelength of 526 nm in ZnS: Cu is associated with radiation in the scattering of carriers from the levels of sulfur vacancies, and a maximum of 556 nm with carrier scattering from the conduction band at defect levels of copper t_2. A maximum with a wavelength of 449 nm in ZnS: Mn is related to scattering by lattice defects (vacancies) of zinc, 566 nm to scattering by Mn^{2+} ions embedded in the ZnS lattice instead of zinc, and a maximum of 520 nm is associated with radiation in the scattering of carriers from vacancy levels sulfur at defective levels of manganese t_2.

Three luminescence bands with maxima of 451, 518, and 542 nm can be distinguished on the electroluminescence spectra of samples doped with Cu; Cl. The emission bands with wavelengths of 451 and 518 nm are associated with the recombination of charge carriers on sulfur and zinc vacancies, respectively, whereas a band with a wavelength of 542 nm is most likely associated with the scattering of carriers from the conduction band at defective levels of copper t_2. Chlorine near the conduction band forms an additional level on which there is an excessive number of electrons, and transitions from a given level to t_2 level of copper substantially increase the intensity of the glow. An increase in the concentration and pore diameter of the substrate matrix leads to an increase in the radiation intensity. This is probably due to the increased concentration of electric field strength on nanostructures, which leads to a more intensive emission of light quanta on them. No major changes in the emission spectra from the substrate material were detected. Two luminescence bands with maxima of 530 and 590 nm can be distinguished on the spectra of samples doped with Mn; Cl, which corresponds to self-activated ZnS luminescence and radiation on Mn^{2+} ions, respectively. The observed shifts of the maxima can be related both by the difference in the concentrations of the main ligand and by the presence of a coactivator.

All the tasks of the project, adjusted in accordance with the obtained intermediate results, are fulfilled. But there are a number of problems, the solution of which can significantly increase the functional characteristics of the materials being developed. So, in the process of modeling, we always took as one of the initial conditions one time, and precipitation in the presence of neighboring pores was not carried out.

It was also found that the optimal pore diameter for obtaining samples with the maximum electroluminescence is 80 nm. But the dependence of structure and properties on the thickness of the boundaries between individual pores as well as on the thickness of the matrix itself is not established. Also, experiments on deposition on substrates at different angles of inclination with respect to the incident molecular beam were not planned or carried out, which may be important from the point of view of maximum pore filling.

An additional analysis of publications over the past 3 years has shown that the achievements in the field of materials science have made it possible to develop ELS based on electrophosphors with quantum structures based on zinc sulfide, on various flexible polymer bases, and also the development of light emitting diodes (LEDs) working on the principles of electroluminescence. Researchers are discovering new mechanisms for the appearance of electroluminescence; their experimental and theoretical justification is being carried out. The researchers are shifting their interest in reducing the thickness of nanostructured phosphor layers to create electroluminescent DC devices. Separation of charges in them can be performed using inorganic hole conductors, for example, NiO_x.

FUTURE RESEARCH

Thus in the future, it is proposed to proceed to the excitation of electroluminescence by direct current with small voltages (up to 10 V), which requires the development of other principles for the formation of a light-emitting device. The electroluminescent layers formed are porous alumina with a thickness of 150–200 nm with a phosphor deposited in the pores. For the injection of holes, a layer of nickel oxide deposited by magnetron sputtering will be used.

To achieve this goal, problems of modeling the growth processes of nanostructures of doped zinc sulfide will be solved in the matrix approximation with pore sizes from 40 to 100 nm.

It is necessary to develop the method of the synthesis of ultrathin (up to 200 nm) porous alumina films on the aluminum surface with removal of the buffer layer at the porous interface matrix/aluminum.

Also, it is necessary to develop the method of precipitation of zinc sulfide doped with copper and manganese ions with chlorine salt consolidation, which ensures full filling of the matrix pores.

It is necessary to develop the Magne technique thin deposition of thin films of nickel oxide as a hole-conducting layer.

It is necessary to investigate their structural, electronic, and optical properties.

It is necessary to investigate the formation of light-emitting devices using matrices with various geometric parameters of the porous structure and thickness, as well as the composition of the luminescent layer.

It is necessary to study of processes occurring at the boundaries of the sections of the layers.

Ultimately, the main task is to identify the relationship between the structural characteristics of the matrix and the phosphor, and the ELS emission characteristics.

CHAPTER 1

Brief Overview of the Literature on Nanostructural Electroluminescent Light Sources

ABSTRACT

Light-emitting electroluminescent panels based on powders and thin films of zinc sulfide (ZnS) electrophosphors are widely used, mainly, in symbol information displays, illumination panels of advertising structures. Modern methods of materials synthesis, such as electrochemical deposition, ink printing, chemical vapor deposition, variations of vacuum-thermal deposition allow forming electroluminescent layers practically on any surface. The research results dedicated to the synthesis of nanostructured materials are also applicable in our case. Phosphors based on 0D (quantum dots, free and stabilized nanoclusters), 1D (nanowires, nanorods, nanotubes, nanoribbons, whiskers), and 2D (thin films up to several hundred nanometers, heterostructures, LB films), 3D (nanoparticles, including those in envelopes, 3D self-organizing arrays of nanoobjects) are actively studied. Luminescent nanoparticles in organic and inorganic dielectric matrixes are of special interest, since the matrix isolation allows not only spatially dividing separate nanoparticles but also protecting them from the environment.

This chapter contains a brief history of the problem, description of electroluminescent light sources (ELS), and electroluminescence mechanism in ZnS ELS and review of phosphors based on impurity-doped ZnS nanostructures.

1.1 A BRIEF HISTORY OF THE PROBLEM

Luminescence in the zinc sulfide (ZnS) films obtained by sublimation method was first discovered by Theodore Sidot in 1866. Its emergence was explained by the presence of copper admixture during the production process.[1] The main reasons and mechanism of luminescence were investigated by Fischer in 1960s. He explained it by the appearance of needle-like defects of Cu_xS-type copper sulfides formed in the process of phase transition of ZnS from wurtzite to cubic phase under cooling.[2]

The recombination of tunneled electrons and holes takes place in the regions close to Cu_xS that demonstrates the predominantly defective type of emission during luminescence.

The investigation of electroluminescence of materials based on ZnS was initiated by Desterio in the 1930s,[3] but the use of fine-crystalline powders or thin films of ZnS in electroluminescent devices with large area of luminous surfaces became most widespread only in the 1980s.[4,5]

The electroluminescence phenomenon is of interest because it lies at the intersection of the whole number of chapters of physics and engineering, such as optic and electric properties of solids, physics, and chemistry of wide-band-gap compounds, electronics, and illumination engineering. The development of new trend in electronics—optoelectronics is also connected with the appearance of electroluminescent emitters.

Electroluminescent emitters have high brightness (10^3 cd/m^2), sufficient service life (10^4 h), and fast response (1 ns). Emitters have various electric and illumination engineering parameters, some types of emitters are well-compatible with circuits with semiconductor devices.

The application of emitters together with photoelectric detectors allows enhancing and converting light signals, as well as electric isolation of different blocks of devices. The general tendency of transition from vacuum and gas-discharge devices to solid-state ones, thus, also befell light sources.[6,7]

In the early 1970s, the successes in phosphor film production technology and the development of structures in which the phosphor film about 1 μm thick was placed between two layers of dielectric allowed obtaining emitters actuated by AC voltage and possessing not only high brightness (up to 9000 cd/m^2) but also longer service life (about 10,000 h) that immediately arouse the interest to such emitters.[8]

It was found that thin-film electroluminescent sources (TFELS) have a number of advantages in comparison with emitters based on powder phosphors.

Among such advantages we can point out not only high brightness, stability of thin-film samples, but also (that is especially important to apply TFELS as the basis to produce flat television and other screens) increased K-rating factor of volt-brightness characteristic (over 30 in comparison with 3–4 for powder light sources) and higher resolution.

The production of TFELS with hysteresis of volt-ampere characteristic allowed developing flat screens with memory.

Every year, hundreds of papers on the materials for electroluminescent light sources (ELS) are published, specialized conferences and symposia are held.

The literature analysis demonstrates that at present the main areas of research in the field of electroluminescent materials are: the obtaining of new, more effective phosphors (including based on organic compounds),[8–10] and the attempt to increase functional characteristics of phosphors based on ZnS.[11–14]

1.2 ELECTROLUMINESCENT LIGHT SOURCES

The typical structures of TFELS both on AC and DC voltage are demonstrated in Figures 1.1 and 1.2. The transparent electrode *2* is deposited to the template *1*, with dielectric layers *3*, *5*, phosphor film *4* and second electrode *6* over it. Glass is used as the substrate.

Wide-band-gap semiconductor compounds: SnO_2; SnO_2:Sb; In_2O_3, etc. are usually used as transparent electrodes.

Transparent metal electrodes ($In_2O_3 \cdot SnO_2$ alloy) can also be used to obtain the emission interference (in this case it is possible to vary the glow color of TFELS with one and the same phosphor), decrease in the electrode resistance, improvement of heat removal from the structure.

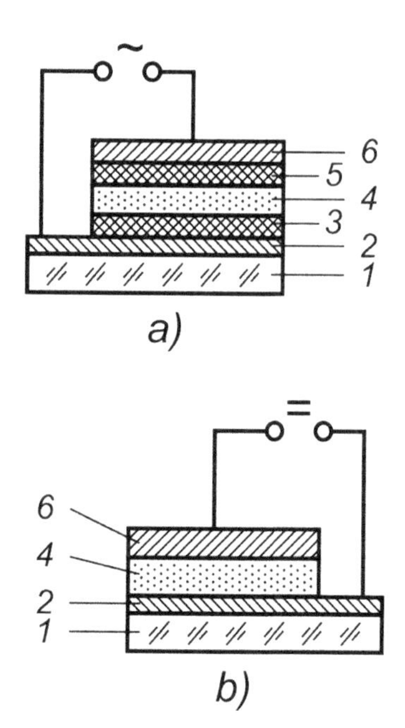

FIGURE 1.1 Layout diagram of TFELS operating on AC (a) and DC (b) voltage.

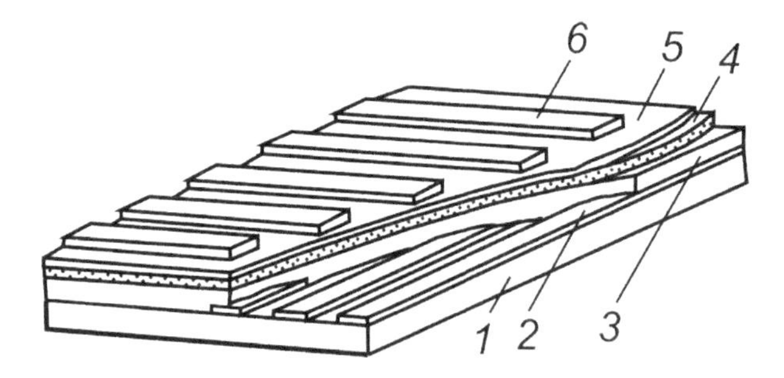

FIGURE 1.2 Diagram of thin-film electroluminescent screen.

The application of dielectric layers in TFELS was caused, first of all, by the necessity to decrease the possibility of catastrophic breakdown of the whole structure.

However, it was found out that these layers play an important part in the mechanism of samples actuating, what was especially clearly noticeable for TFELS with two dielectric layers. Apart from the fact that such layers decrease the possibility of phosphor film decomposition limiting the current going through it, the interface "phosphor–dielectric" is an effective source of electrons accelerated later by the electric field.

When creating TFELS, the highest brightness at the sufficient reserve of electric strength is achieved when the ratio between phosphor thickness d_1 and each of these two dielectric layers d_2 is d_2=0.4d_1. Usually, d_1≈0.5–1.5 μm, d_2≈0.2–0.5 μm, at the same time, the thicknesses of two dielectric layers can differ.

The methods used when creating TFELS are largely similar to those usually used in microelectronics to produce film structures.[15] The general requirement to production, provision of uniform properties on the significant area, is a rather complex technological task.

1.3 ELECTROLUMINESCENCE MECHANISM IN ZNS ELS

The largest part of electric and optic characteristics of AC voltage ELSs based on impurity-doped ZnS can be explained with the help of the following process model. Ionization > (thermal, by electric field, thermal-field) of impurity and defects of luminescent layer, as well as states at the interface dielectric/conductor take place in the presence of strong electric field (W=10^5–10^6 V/cm) (Fig. 1.3).

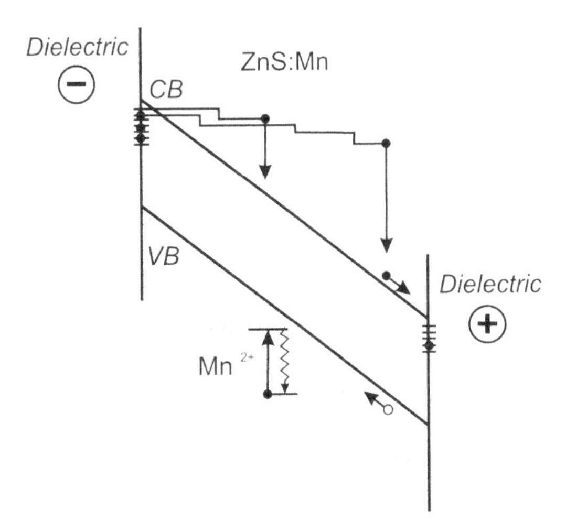

FIGURE 1.3 Excitation of electroluminescence in the structure with two dielectrics. Position of manganese levels by the energy scale is demonstrated conditionally.

Freed electrons accelerated by the field rush from the first electrode (cathode) to the second one positively charged at that moment. They excite impurity atoms (e.g., manganese) and crystalline lattice of ZnS on their way. At the same time, holes emerging during the lattice ionization move to the cathode. By the end of alternation of alternating electric field action, the negative charge is located on the entrapping levels at the "phosphor—dielectric" boundary. The positive spatial charge conditioned by the ionized centers and deep traps, having entrapped the holes, is localized at the cathode side of the film.[4]

During the second alternation, the electric field of the spatial charge folds with the external one considerably enhancing it. Electrons freed from the entrapping levels rush back to the first electrode, being the anode at the first alternation. At the same time, ionization of different centers and atoms of the lattice itself takes place, as well as excitation of activator ions.

Part of electrons recombine with the centers ionized during the first alteration, the rest are entrapped by the states at the interface "phosphor–dielectric," this time from the side of the first electrode. After the end of the second alteration, the process is repeated.

The basis of the model is the assumption on the availability of considerable number of traps both at the boundary "phosphor–dielectric" (density of such states equals approximately 10^{16}–10^{17} m^{-2}) and in the volume of the conductor layer. At the same time, it is considered that the field intensity necessary for tunnel penetration of electrons into the phosphor layers is sufficient for the beginning of shock processes in crystalline phosphor that usually takes place in ELS based on ZnS.

The glow occurs because of the excitation of the impurities ions with hot electrons. In ELS based on ZnS phosphors the glow has intercenter nature; the mechanism of impurity excitation is shock.

1.4 PHOSPHORS BASED ON DOPED ZnS NANOSTRUCTURES

As mentioned before, powder and film phosphors based on ZnS have been successfully investigated for a long time, fundamental mechanisms of both impurity and own luminescence have been studied and they are applied in production of commercial products. But because of the discovery of the influence of size factors onto the properties of semiconductor luminescent materials, the interest to them decreases and they are the objects of intensive investigations mainly targeted at the increase in consumer properties of devices based on them.

A lot of survey publications come out on the back of this interest, the paper by Fang et al. being the most interesting among them,[16] disclosing the main properties and perspectives of phosphors based on ZnS nanostructures not only as applicable to electroluminescence but also to cathode luminescence, chemiluminescence, and electrochemiluminescence.

It should be pointed out that quite a lot of time passed since the review publication of the above survey and investigations significantly progressed both in the improvement of synthesis methods and production of materials with more eminent properties.

The schematic view of ELS obtained by ink printing the emission layer with the content of ZnS:Mn nanoparticles is given in Figure 1.4.[17]

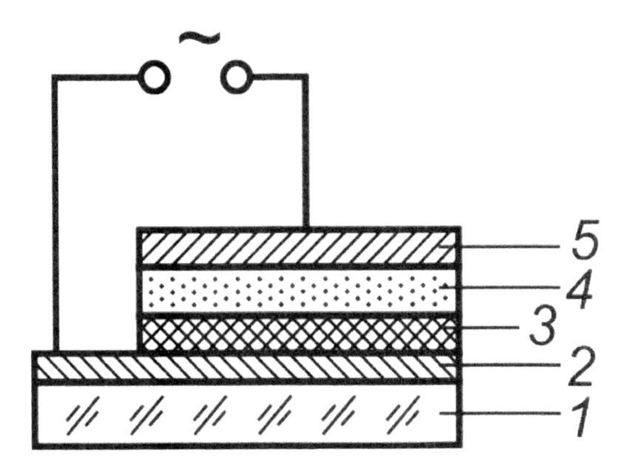

FIGURE 1.4 Schematic view of ELS obtained by printing the emission layer with the content of ZnS:Mn nanoparticles: 1—glass substrate, 2—transparent electrode (ZnO:Al), 3—isolating layer (Si_3N_4), 4—ink-printed ZnS:Mn, 5—aluminum contact.
Source: Modified with permission from Ref. [17].

The transparent electrode based on ZnO:Al, isolating layer Si_3N_4, emission layer of ZnS:Mn nanoparticles, and upper aluminum electrode are formed on the glass template. ZnS:Mn nanoparticles were obtained by the method of chemical codeposition. First, the solution containing ions Zn^{2+} and Mn^{2+} was prepared by diluting $Zn(CH_3COO)_2{\cdot}2H_2O$ and $Mn(CH_3COO)_2{\cdot}4H_2O$ in distilled water.

The molar ratio $[Zn^{2+}]:[Mn^{2+}]$ was 1:0.1. Then, the solution containing ions S^{2-} was made by diluting $Na_2S \cdot 9H_2O$ in distilled water at 80°C. Finally, the solutions $[Zn^{2+}]:[Mn^{2+}]$ and $[S^{2-}]$ were mixed in ultrasonic bath at 55°C. To obtain nanoparticles with stoichiometric composition, it was necessary to provide the molar ratio $[Zn^{2+}]:[S^{2-}]$ equaled to 1:1.67.

The colloid particles of ZnS:Mn were separated by centrifugation, washed to remove residual ions, and dried. The inks based on rubber diluted in organic solvent and dispersed ZnS:Mn nanoparticles were prepared to print the emission layer. After the deposition, the layer was dried at 130°C.

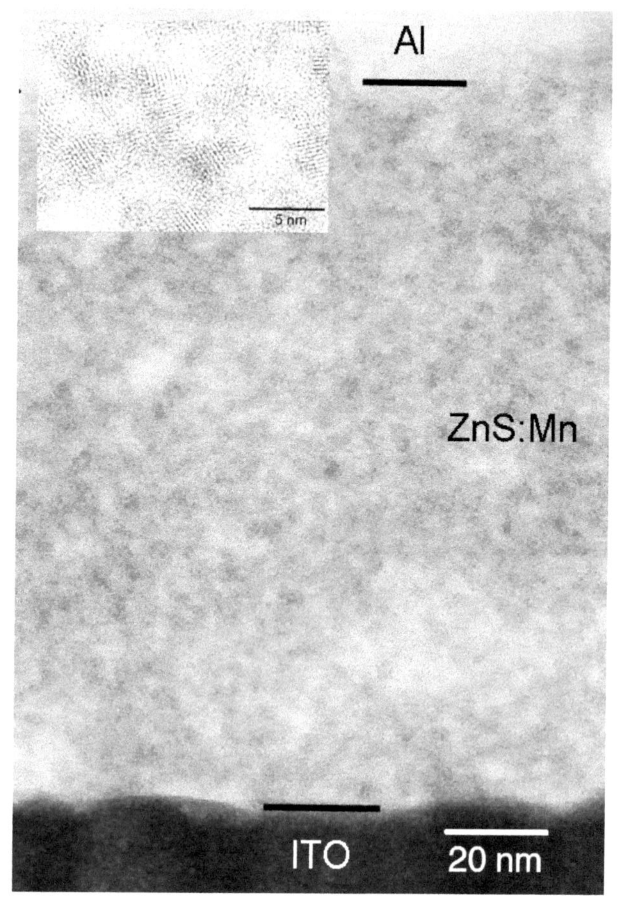

FIGURE 1.5 Image of ELS structure with the layers Al/emission layer/ITO, insert represents magnified image of the phosphor layer containing ZnS:Mn nanoparticles.
Source: Modified with permission from Ref. [17].

It is seen in the high-resolution electron microscopy pictures (Fig. 1.5.) that ZnS nanoparticles with the average size of 3.4 nm are uniformly spread in the film volume. X-ray diffraction investigations demonstrated cubic crystalline structure of nanoparticles.

The estimation of the band gap showed the value of 3.85 eV, and for pure ZnS it equals 3.7 eV. The authors link the increase of this value with the influence of quantum effects because of the littleness of nanoparticle sizes in the film in comparison with the average exciton radius (4.4 nm).[18]

Electroluminescence was investigated using AC voltage with the frequency of 5 kHz. For further investigations, the authors used the voltage of 150 V, that is, by 45 V higher that the voltage of emission emergence. ELS obtained possess luminescence in the orange region of spectrum with the wavelength of 590 nm caused by the transitions $^4T_1 \rightarrow {}^6A_1$ of ions Mn^{2+}.[19]

As discussed by Adachi et al.,[13] the excitation of direct shock under the action of "hot" electron is the predominant mechanism of electroluminescence in ZnS:Mn nanoparticles. As a result, the electron-phonon interaction in ZnS nanoparticles is important to increase electron energy in strong electric fields.

With the increase in particle size, the emission brightness intensity goes down, which is connected with the increase in electron–phonon interaction and, consequently, decrease in the energy of "hot" electrons and emission brightness.[14]

The paper[20] is another example of nanoparticles application in the emission layer of ELS, where it is proposed to use the alternating layers of ZnS (12 nm) and ZnSe/ZnS:Mn/ZnS nanoparticles (30 nm) (Fig. 1.6). The authors demonstrated that the interface between the film of ZnS and nanoparticles is the active region of phosphor.

The number of alternating layers also playsa great role (Fig. 1.7a); the more are the layers, the more is the emission brightness. Three main maxima are seen in the electroluminescence spectrum in Figure 1.7c. The broad background emission in the interval from 450 nm to near IR-region is related to the electric excitation of defective states of ZnSe, which can be formed in the process of ZnS layer deposition.

This also confirms the fact that this background is missing in photoluminescence spectra. The maximum at the wavelength of 590 nm, the same as in the first example,[17] is related to the transition $^4T_1 \rightarrow {}^6A_1$ of Mn^{2+} ions.

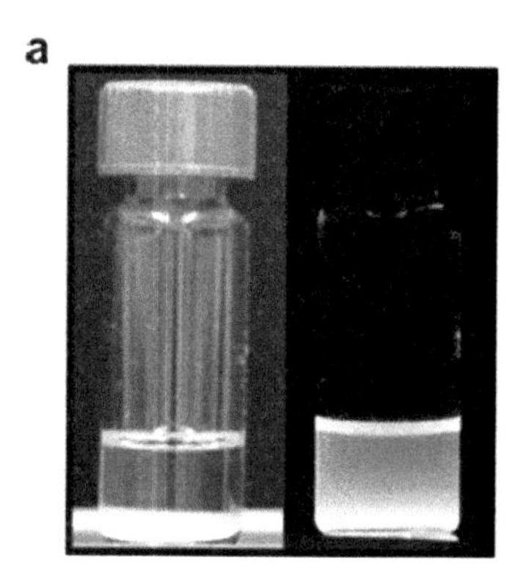

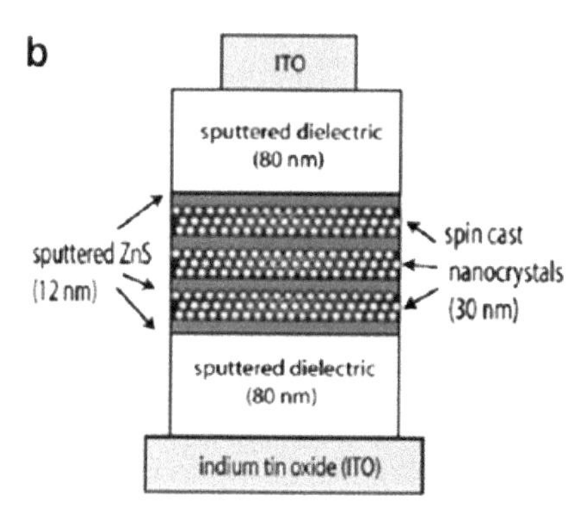

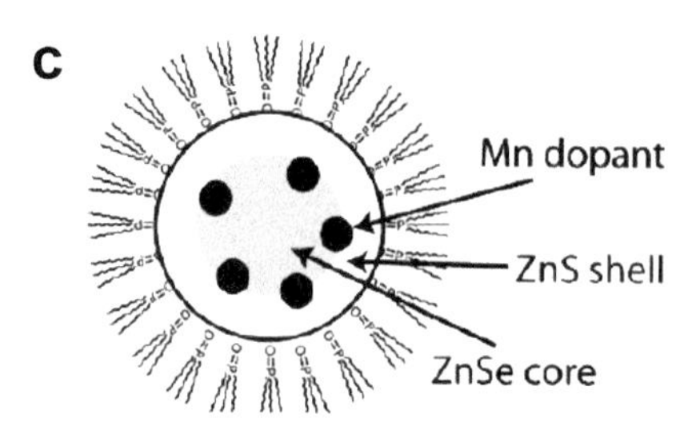

FIGURE 1.6 Photograph of the solution of ZnSe/ZnS:Mn/ZnS crystals in chlorophorm in the daylight and darkness (a, b). Schematic view of ELS based on ZnS layers (12 nm) and ZnSe/ZnS:Mn/ZnS nanoparticles (c).
Source: Reprinted with permission from Ref. [20]. Copyright 2009 American Chemical Society.

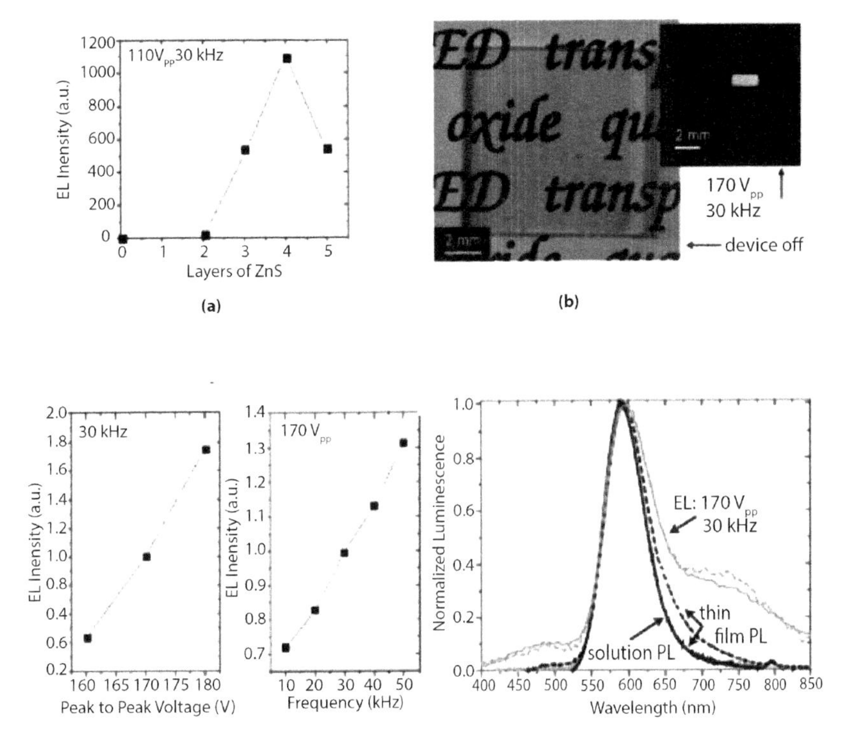

FIGURE 1.7 (See color insert.) (a) Graph of dependence of electroluminescence intensity on the number of alternating layers of ZnS:Mn. (b) Photograph of ELS sample without the current supply and at electroluminescence excitation by the field with the frequency of 30 kHz and voltage of 170 V; (c) dependences of electroluminescence intensity on the voltage (at the left) and frequency (at the right) of the excitation electric field; (d) spectra of electroluminescence samples with four alternating layers of ZnS with dielectric layers of Al_2O_3 (solid orange line) and HfO_2 (dashed orange line) and luminescence spectra of ZnS:Mn particles in the form of suspension in chlorophorm (dashed black line) and in the form of film in the structure ITO/Al_2O_3/ZnS:Mn/30 (solid black line).
Source: Reprinted with permission from Ref. [20]. Copyright 2009 American Chemical Society.

Figure 1.7b shows the emission intensity significantly depends on the frequency and voltage of applied AC electric field: The electroluminescence intensity goes up with voltage increase at fixed frequency and frequency increase at fixed voltage.

The authors also point out that a great role is played by the material of intermediary dielectric layers: The use of Al_2O_3 is more preferable than HfO_2 since the electroluminescence intensity is two times higher.

The paper by Mishra et al.[21] presents the comparison of radiating properties of the samples with ZnS:Cu layers based on microcrystalline particles of ZnS:Cu (the authors call them as "bulk") and nanoparticles.

Microparticles for "bulk" samples were obtained by annealing the well-mixed mixture of ZnS and Cu in the pressure-tight quartz tube at 800°C, and nanoparticles as a result of chemical reaction in the mixture of Zn-acetate and Cu-acetate in thiourea at 110°C. ELS schematically demonstrated in Figure 1.8a were produced to investigate the electroluminescence.

The 10 μm thick luminescent layer produced by mixing microparticles and nanoparticles, respectively, in the binder was applied onto the polyethylene template with ITO layer. Then, the dielectric layer of barium titanate and contact from silver paste were applied.

The spectra of "bulk" ZnS:Cu sample luminescence obtained at 10-kHz frequency of excitation current are given in Figure 1.8b. It is seen that the radiation intensity goes up with the voltage increase.

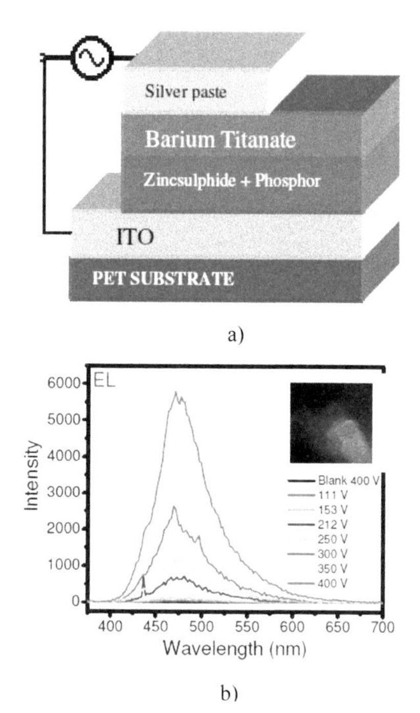

FIGURE 1.8 (See color insert.) Scheme of ELS with working layers based on ZnS:Cu micro- and nanoparticles (a) and electroluminescence spectra of the sample with phosphor based on microparticles (b).

Sources: Reprinted with permission from Ref. [21]. Copyright 2015 AIP Publishing.

The efficiency was 1.5%. For the samples with nanoparticles, the electroluminescence was not observed even with the voltage of 400 V. The authors conclude that such difference in properties can be related to the structure of ZnS:Cu micro- and nanoparticles.

The structure, particle morphology, and photoluminescent properties of Mn^{2+}-doped ZnS nanoparticles are reported by Raleaooa et al.[22]

The chemical reaction for the preparation of ZnS nanoparticles is presented by eq (1.1):

$$ZnCl_2(s) + Na_2S(s) \rightarrow ZnS(s) + 2Na+ (aq) + 2Cl- (aq) \qquad (1.1)$$

ZnS:Mn^{2+} nanopowder samples were prepared using a sol–gel method by dissolution of $Zn(CH_3COO)_2{\cdot}2H_2O$ (2.00 g, 0.011 mol) in 35 mL of deionized water. Na_2S (0.78 g, 0.01 mol) increased the pH of the solution to 12 around which the precipitation of Zn^{2+} was favored. This is a condition that is just right to drive a reaction rate that allows the addition of adopant with $Mn(CH_3COO)_2$ concentrations (x = 0.0, 0.5, 1.5, and 4.0 mol%). Equation (1.2) describes the reaction that led to the formation of ZnS:Mn^{2+}.

$$Zn_{1-x}(CH_3COO)_2{\cdot}2H_2O\ (s) + 2Na_2S\ (s) + Mn_x(CH_3OO)_2\ (s) \rightarrow ZnS_{1-x}Mn_x\ (s) + 2H_2O(l) + +4CH_3COO-(aq) + 4Na+ (aq) \qquad (1.2)$$

The X-ray powder diffraction studies showed that incorporation of Mn^{2+} did not cause any change to the crystallized cubic (sphalerite) phase of ZnS but it reduced crystallinity because of increased incoherent scattering, resulting in increased broadening of the diffraction peaks.

The average crystallite size calculated from the broadening of the X-ray diffraction peaks using Scherrer equation was ~2 nm, implying that our particles were highly quantum confined.

The field emission scanning electron microscopy confirmed that the undoped ZnS powders consisted mainly of an agglomeration of spherical particles, whereas the Mn^{2+}-doped powders were consistent of particles with irregular shapes.

Authors obtained a deconvoluted photoluminescence (PL) emission spectrum of ZnS nanoparticle phosphor using the Gaussian function. The spectrum was fitted into four peaks located at 379, 417, 433, and 481 nm, respectively. The peaks at 379, 417, and 433 nm are respectively assigned to interstitial sulphur (I_S), zinc interstitial (I_{Zn}), and recombination of an electron from sulphur vacancy donor level (V_S) with holes in the valence band, whereas the tail of the PL spectrum, extending beyond 500 nm, is attributed to zinc vacancies (V_{Zn}).[23]

For the ZnS doped with Mn^{2+} nanoparticles, the orange emission peak centered at 600 nm was observed in addition to the blue emission coming from the ZnS nanoparticles. The emission at 600 nm is because of electronic transitions from the excited state $^4T_1$ to the ground state $^6A_1$ within the 3d shell of Mn^{2+}. The peak intensity of this emission is shown to increase with Mn^{2+} concentration, whereas the blue emissions from the ZnS nanoparticles were decreasing.

According to Bhargava et al.,[18] when Mn^{2+} ions substitute Zn^{2+} cationic sites in ZnS, the mixing of s-p electrons of the ZnS host with the 3d electrons of Mn^{2+} causes strong hybridization and makes the forbidden transition of $^4T_1$–$^6A_1$ to be partially allowed, giving rise to the orange–red emission at 600 nm. The inset show the snapshots of the emission colors associated with different spectra recorded when the powders were excited using 365 nm ultraviolet lamp.

Upon UV excitation at 325 nm, electrons from the valence band are promoted to the conduction band of the ZnS host lattice, followed by radiative relaxation that involves multiple pathways such as the excitonic recombination and defects trap states creating additional strong emissions of light. Furthermore, the Mn^{2+} dopant ion can also be excited according to two possible mechanisms.

The following scheme is derived directly from the PL spectra and represents typical transitions in Zn:Mn phosphors. This includes excitation (339 nm) of electrons from the valence band to the conduction band, recombination of the electron from the conduction band with a hole trapped in the I_S giving emission at 379 nm, recombination of the electron from the conduction band with the hole trapped in the V_{Zn} giving emission at 481 nm, recombination of detrapped electron from the V_S with the hole in the valence band giving emission at 433 nm and Mn^{2+} atomic $^4T_1 \rightarrow {}^6A_1$ transition giving emission at 600 nm (Fig. 1.9).

It should be pointed out that the majority of modern publications are dedicated to ELS based on microcrystals or films of impurity-doped ZnS obtained by different, mainly, chemical methods, whereas in phosphors based on nanoparticles of alloyed ZnS, mainly UV-PL is investigated.[16,22]

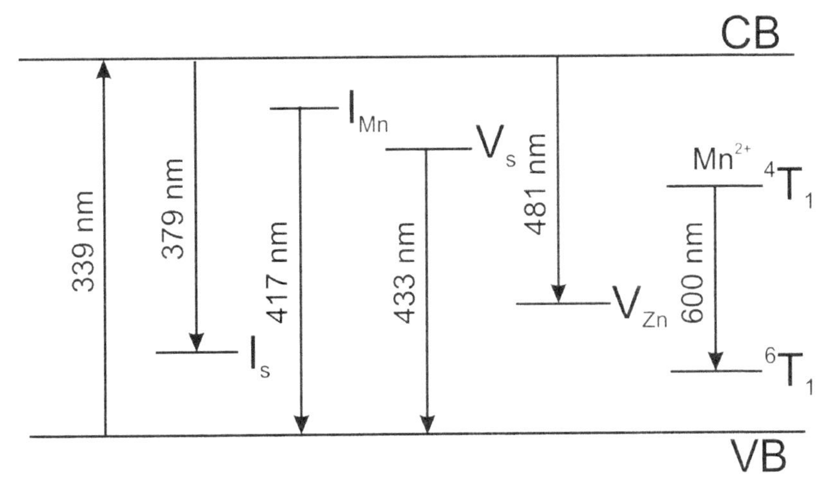

FIGURE 1.9 Schematic diagram of energy levels in ZnS:Mn^{2+} nanoparticle phosphors corresponding to photoluminescence spectra.
Source: Modified with permission from Ref. [22].

This raises the problem of creating and developing such methods, which would allow obtaining nanosized particles of phosphors by physical methods of deposition, such as thermal, magnetron, etc.

As discussed above, as a rule, two dielectric buffer layers are formed when creating ELS.[6] They are necessary to accumulate (similar to capacitor) charge near the interface "dielectric/conductor" because of which high intensity of the electric field is produced between the conducting plates. The more developed is the dielectric surface, the higher is the intensity of the field created.[24]

Besides, the dielectric playing the role of the carrier of phosphor nanoparticles can protect the material from the environment influence and thus improve the operational characteristics of ELS. It should be also pointed out that the ordered array of phosphor material nanostructures with uniform size and shape can be formed by using the porous anodic aluminum oxide.

This allows presenting each nanostructure as a separate light source. In this case, the coherent addition of radiation from each source will result in significant increase in the light intensity.[25]

KEYWORDS

- **electroluminescence basis**
- **electroluminescent nanoparticles**
- **electroluminescent light sources**
- **ZnS:Cu**
- **ZnS:Mn**

REFERENCES

1. Yen, W. M.; Shionoya, S.; Yamamoto, H. *Phosphor Handbook*, 2nd ed; CRC Press, Taylor & Francis Group: Boca Raton FL, 2007; p 625.
2. Fischer, A. G. Electroluminescent Lines in ZnS Powder Particles. *J. Electrochem. Soc.* **1963,** *110*, 733–748.
3. Destriau, G. Experimental Studies on the Action of an Electric Field on Phosphorescent Sulfides. *J. Chem. Phys.* **1963,** *33*, 620–626.
4. Fock, M, Ed. *Applied Electroluminescence*; Soviet Radio: Moscow: 1974; p 416 (in Russian).
5. Gutorov, M. *The Basis of Light Technology and Light Sources*; Energoatomizdat: Moscow, 1983; p 384 (in Russian); Aizenberg, Yu., Ed. *A Background Book in Light Technology*; Energoatomizdat: Moscow, 1983; pp 472 (in Russian).
6. Vereschagin, I; Kovalev, B.; Kosyachenko, L.; Kokin, S. *Electroluminescent Light Sources*: Energoatomizdat: Moscow, 1990; pp 168 (in Russian).
7. Suzuki, Ch.; Inoguchr, T.; Mito, S. Thin Film EL Displays. *J. Inform. Display* **1977,** *13*, 14–19.
8. Sheats, J. R.; Antoniadis, H.; Hueschen, M.; Leonard, W.; Miller, J.; Moon, R.; Roitman, D.; Stocking, A. Organic Electroluminescent Devices. *Science* **1996,** *273*, 884–888.
9. Chen, S.; Wei, J.; Wang, K.; Wang, C.; Chen, D.; Liu, Y.; Wang, Y. Constructing High-performance Blue, Yellow and Red Electroluminescent Devices Based on a Class of Multifunctional Organic Materials. *J. Mater. Chem. C* **2013,** *1*, 6594–6602.
10. Rahman, N. H. A.; Manut, A.; Rusop, M. Review on Electroluminescence Behaviour of Organic Light Emitting Diode. *Adv. Mater. Res.* **2014,** *832*, 455–459.
11. Ummartyotin, S.; Bunnak, N.; Juntaro, J.; Sain, M.; Manuspiya, H. Synthesis and Luminescence Properties of ZnS and Metal (Mn, Cu)-doped-ZnS Ceramic Powder. *Solid State Sci.* **2012,** *14*, 299–304.
12. Li, Y.; Zhou, S.; Chen, Z.; Yang, Y.; Chen, N.; Du, G. Luminescence Properties of Br-doped ZnS Nanoparticles Synthesized by a Low Temperature Solid-state Reaction Method. *Ceramics International* **2013,** *39*, 5521–5525.

13. Adachi, D.; Takei, K.; Toyama, T.; Okamoto, H. Excitation Mechanism of Luminescence Centers in Nanostructured ZnS:Tb, F Thin-film Electroluminescent Devices. *Jpn. J. Appl. Phys.* **2007,** *47*, 83–86.
14. Chan, W.; Sammynaiken, R.; Huang, Y.; Malm, J. O.; Wallenberg, R.; Bovin, J. O. Crystal Field, Phonon Coupling and Emission Shift of $Mn2^{+}$ in ZnS:Mn Nanoparticles. *J. Appl. Phys.* **2001,** *89*, 1120–1129.
15. Messel, L; Glang, R., Eds. *Handbook of Thin Film Technology*; McGraw Hill Hook Company, New York, USA, 1970; p 800.
16. Fang, X.; Zhai, T.; Gautam, U.K.; Li, L.; Wu, L.; Bando, Y.; Goldberg, D. ZnS Nanostructures: From Synthesis to Applications. *Prog. Mater. Sci.* **2011,** *56*, 175–287.
17. Toyama, T.; Hama, T.; Adachi, D.; Nakashizu, Y.; Okamoto, H. An Electroluminescence Device for Printable Electronics Using Coprecipitated ZnS:Mn Nanocrystal Ink. *Nanotechnology* **2009,** *20*, 055203 (5pp).
18. Bhargava, R. N.; Gallagher, D.; Hong, X.; Nurmikko, A. Optical Properties of Manganese-doped Nanocrystals of ZnS. *Phys. Rev. Lett.* **1994,** *72*, 416–419.
19. Bol, A. A.; Meijerink, A. Long-lived Mn^{2+} Emission in Nanocrystalline ZnS:Mn^{2+}. *Phys. Rev. B* **1998,** *58*, R15997–R1600.
20. Wood, V.; Halpert, J. E.; Panzer, M. J.; Bawendi, M. G.; Bulovic, V. Alternating Current Driven Electroluminescence from ZnSe/ZnS:Mn/ZnS Nanocrystals. *Nano. Lett.* **2009,** *9*, 2367–2371.
21. Mishra, R. K.; Satya, K.; Patel, D. K.; Ramachandra, R. K.; Sudarsan, V.; Vatsa, R. K. Luminescence from ZnS: Bulk vs Nano. *AIP Conf. Proc.* **2015,** *1665*, p 050154.
22. Raleaooa, P. V.; Roodt, A.; Mhlongo, G. G.; Motaung, D. E.; Ntwaeaborw, O. M. Analysis of the Structure, Particle Morphology and Photoluminescent Properties of ZnS:Mn^{2+} Nanoparticulate Phosphors. *Optik* **2018,** *153*, 31–42.
23. Karar, N.; Singh, F.; Mehta, B.R. Structure and Photoluminescence Studies on ZnS:Mn Nanoparticles. *J. Appl. Phys.* **2004,** *95*, 656–660.
24. Yang, Y.; Huang, J. M.; Liu, S. Y.; Shen, J. C. Preparation, Characterization and Electroluminescence of ZnS Nanocrystals in a Polymer Matrix. *J. Mater. Chem.* **1997,** *7*, 131–133.
25. Botez, D.; Scifres, D. R. *Diode Laser Arrays*. Cambridge University Press: Cambridge, UK, 2005; p 468.

CHAPTER 2

Porous Anodic Aluminum Oxide: Structure, Properties, and Application in Semiconductor Technology

ABSTRACT

This chapter is devoted to porous anode aluminum oxide and its structure, properties, and application in semiconductor technology. Types of aluminum oxide films formed during anodizing, structure of anode aluminum oxide films, mechanism of porous structure formation, and influence of anodizing parameters on the structure of anode aluminum oxide film are described in detail in this chapter. In addition, the use of porous aluminum oxide to obtain nanostructures of semiconductors is presented in the end of the chapter.

2.1 TYPES OF ALUMINUM OXIDE FILMS FORMED DURING ANODIZING

The technology of aluminum anodizing and application of alumina films as protective and decorative coatings has had a long history. The first patent for using anodized coatings for corrosion protection of aluminum and its alloys appeared in 1923.[1,2] Quite recently, anodic aluminum oxide has again attracted attention of researchers due to the discovery of the possibility to obtain aluminum oxide films with self-ordered porous structure by two-stage anode oxidizing.[3,4]

This has made the breakthrough in the technology of obtaining porous structures with very narrow distribution of pores by sizes and high geometric anisotropy. The unique porous structure, parameters (diameter, length, and distance between neighboring pores), which can be varied in the process of synthesis allow using films of porous aluminum oxide as inorganic membranes, templating material for the synthesis of nanowires or nanotubes with controllable diameter and high geometric anisotropy, as well as 2D photonic.[5]

At present time, porous aluminum oxide films are most widely applied as matrixes for the synthesis of ordered arrays of anisotropic nanostructures of different composition. The interest to such structures can be explained by the possibility to study fundamental problems (self-organization processes and magnetism in spatial-ordered nanosystems) to solve a wide range of issues related to the creation of high-efficient heterogenic catalysts and to obtain magnetic nanocomposites for information storage devices with extra-high recording density.

On the other hand, the synthesis of anisotropic metal nanostructures is interesting from the point of creating materials with a huge roughness factor and specific surface area that is especially important for producing catalytically active materials.[6] When using Al_2O_3 films as a matrix, it is possible to combine the flexibility of electrochemical methods of obtaining metal catalysts, which allows controlling the properties of electrolytic deposits with the idea of stabilization (fixation) of nanoparticles in the inert matrix of Al_2O_3.

Nanocomposites based on semiconductors in dielectric matrix, matrixes of wide-band-gap, and narrow-band-gap semiconductors are attracting more and more attention. Such structures give the possibility not only to avoid the charge transfer between separate quantum dots with the help of spatial separation of elements but also to protect nanostructures from the environment. The use of matrix isolation based on porous aluminum oxide allows significantly decreasing the efficiency of exciton recombination on surface defects, and, consequently, increasing the quantum yield of nanocomposites luminescence if using them in photovoltaic (light-absorbing and light-emitting) devices. Besides, the matrix availability gives the possibility to significantly decrease the recombination efficiency of electron–hole pairs on semiconductor defects, thus opening vast prospects for designing in nanoelectronics elements (new generation of diodes and switches) based on such systems.

During the anodizing of metal aluminum, depending on the electrolyte composition, barrier and porous films can be formed. In case of aluminum anodizing in neutral solutions with pH = 5–7 (boric acid, ammonium borate and tartrate, phosphate solutions), in which aluminum oxide does not dissolve, nonconductive, nonporous, dense oxide films are formed.[7–10] At the same time, to form porous films of aluminum oxide, the anodizing is carried out in the voltage range from 5 up to 400 V in acid electrolytes, such as solutions of sulfuric,[11] oxalic,[3] phosphoric,[12] and chromic acids.[13]

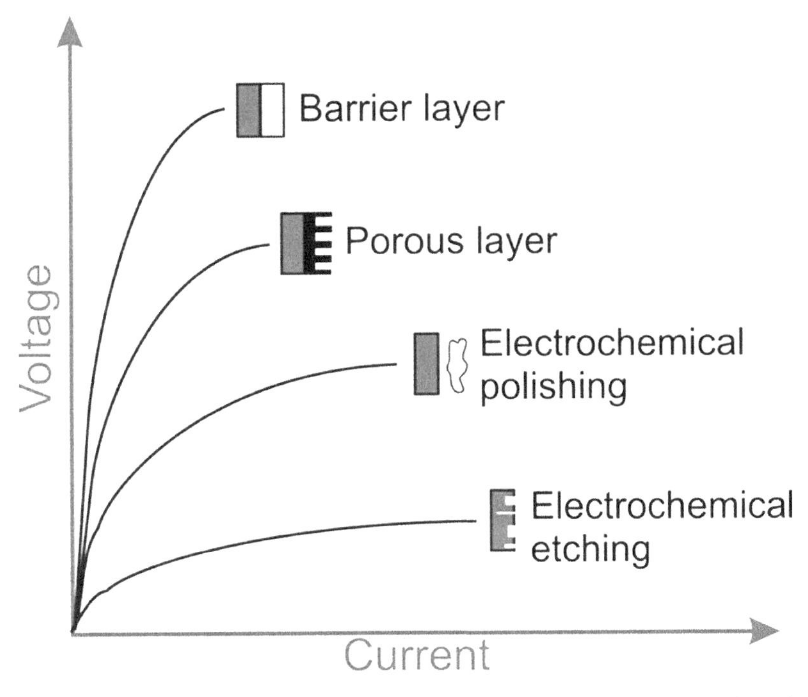

FIGURE 2.1 Processes running during aluminum anodizing in different electrolytes varying the current density and voltage ratio.

Source: Modified from Ref. [14].

In general, the processes during the anode polarization of aluminum with varied current and voltage densities are given in Figure 2.1. Thus, the etching along the crystallographic faces is observed with low voltages and high current densities; the process of electrochemical polishing runs with increased voltage and decreased current density, and further voltage increase results in the formation of porous layer on the surface. Oxide films of barrier type are formed on aluminum surface at high voltages and low current densities.[14]

2.2 STRUCTURE OF ANODIC ALUMINUM OXIDE FILMS

The structure of anodic aluminum oxide porous films can be represented in the form of closely packed hexagonally ordered array of cells with a pore in the center of each cell (Fig. 2.2). The porous structure is characterized by the following parameters: pore diameter (d_p), pore wall thickness (d_w), distance between pore centers (d_{int}), barrier layer thickness (L_{bl}), and oxide film thickness (L)—these parameters of porous structure are given in Figure 2.2.

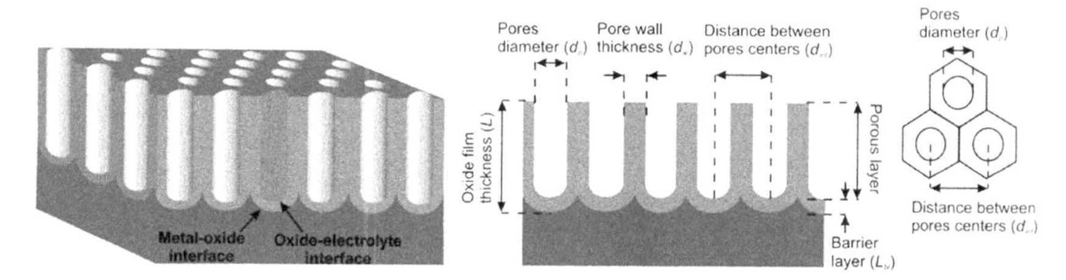

FIGURE 2.2 Schematic representation of the structure of the anodic alumina film and the designation of the main parameters of the porous structure.

Besides, the structure of anode aluminum oxide films is characterized by porosity parameter (ε) calculated as the ratio between pore volume and volume of the whole oxide film:

$$\varepsilon = \frac{V_{pore}}{V_{film}} \tag{2.1}$$

If the pore diameter is constant along the whole film thickness, the porosity of aluminum oxide can be calculated as the ratio between pore area and oxide film area:

$$\varepsilon = \frac{S_{pore}}{S_{film}} \tag{2.2}$$

In the approximation of hexagonal ordering of pores (Fig. 2.2), the formula for porosity calculation is as follows:

$$\varepsilon = \frac{\pi}{2\sqrt{3}} \cdot \left(\frac{d_p}{d_{int}} \right)^2 \tag{2.3}$$

To characterize the porous structure, it is also possible to use such parameter as pore density (number of pores per area unit, N_p), which, in the approximation of hexagonal ordering of pores, can be calculated as follows:

$$N_p = \cdot \left(\frac{\sqrt{3} d_{int}^2}{2} \right)^{-1} \tag{2.4}$$

2.3 MECHANISM OF POROUS STRUCTURE FORMATION

The typical dependence of current density on time during potentiostatic anodizing is given in Figure 2.3. At the first stage of the anodizing process, the exponential decrease in the current density related to the formation of barrier oxide layer is observed (region 1). Then, the dependence of current density on time proceeds through the minimum (region 2), after that it sharply goes up and proceeds through the maximum (region 3) and reaches the constant value (region 4).[15] The current density j_p corresponding to the formation of porous film can be resolved into two components: j_b—current density during the formation of barrier-type film and j_{hp}—hypothetic current density related to the pore formation. The current density j_b is found only by the potential applied, whereas j_{hp} depends on the electrolyte used, temperature at which the oxidation proceeds, and anodizing voltage. It should be pointed out that two modes are distinguished depending on the value of current density during anodizing: the so-called anodizing under "hard" conditions (current density of 30–250 mA/cm^2) and anodizing under "soft" conditions (current density of 5 mA/cm^2).[16]

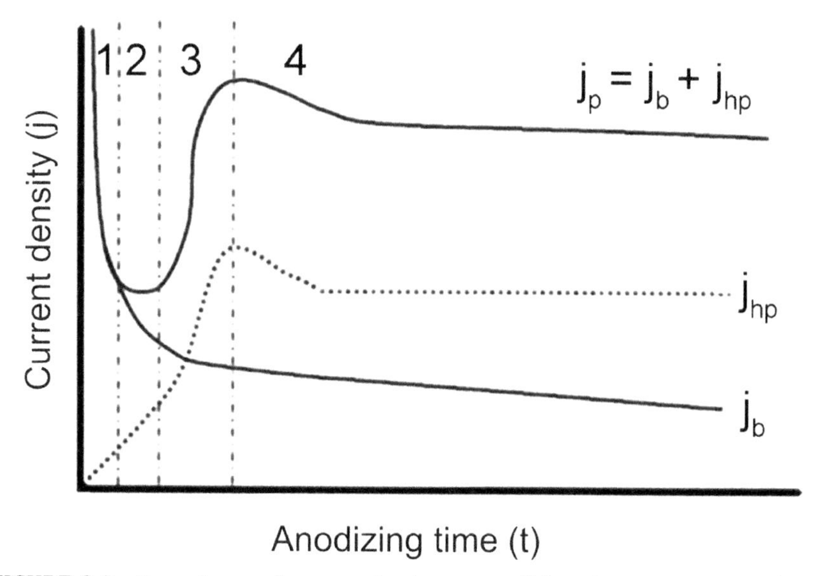

FIGURE 2.3 Dependence of current density on anodizing time at constant voltage; j_b—current density during the barrier-type film formation; j_p—current density during the porous-type film formation; and j_{hp}—hypothetic current density, which represents the difference of j_p.

The processes running at each stage of chronoamperometric dependence are schematically demonstrated in Figure 2.4. At the first oxidation stage, aluminum surface is covered by the barrier layer consisting of nonconductive aluminum oxide ($r = 10^{10}–10^{12}$ Ω cm).[17] The electric field intensity sharply increases in recesses and defects of the oxide film (Stage 2 in Fig. 2.4), resulting in the proceeding of oxide dissolution process both due to the temperature local growth and oxide dissolution under the electric field action (Stage 3 in Fig. 2.4). Aluminum oxide dissolution under the electric field action takes place due to the polarization of Al–O bond with further output/detaching of Al^{3+} ion from the oxide structure.[18] Due to the competition between two neighboring dots of charge drain a part of pores stop growing, as a result, the oxide film starts uniformly growing with constant distance between pore centers.

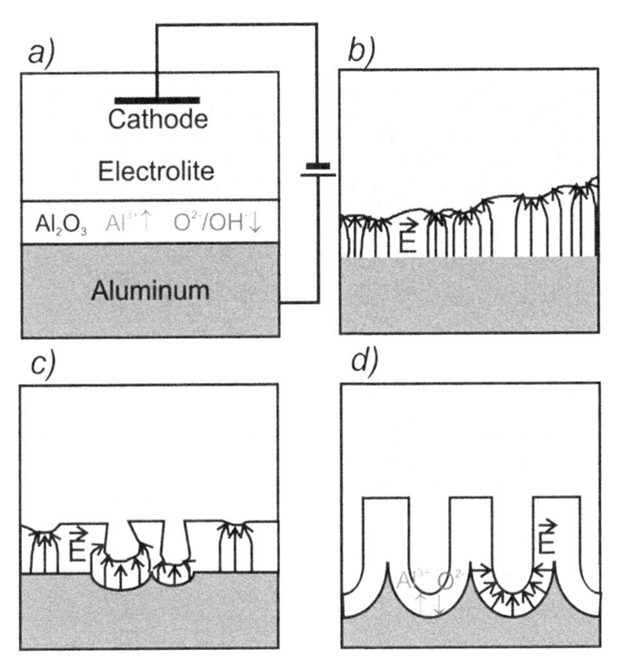

FIGURE 2.4 Schematic view of stages of porous structure formation. Stage 1—barrier layer growth; stage 2—concentration of electric field force lines in recesses of the oxide film; Stage 3—initiation of pores due to the localization of force lines and local overheating; and Stage 4—uniform growth of pores.

2.4 INFLUENCE OF ANODIZATION PARAMETERS ON THE STRUCTURE OF ANODIC ALUMINUM OXIDE FILM

2.4.1 Distance Between Pore Centers

The distance between pore centers is proportional to the voltage during anodization[16,19]:

$$d_{\text{int}} = k \cdot U_{\text{volt}}, \tag{2.5}$$

where *k*—proportionality coefficient and U_{volt}—voltage during anodization.

Coefficient *k* binding the distance between pores and anodization voltage varying in the range between 2.2 and 2.8 nm/V depending on anodization conditions (Fig. 2.5).

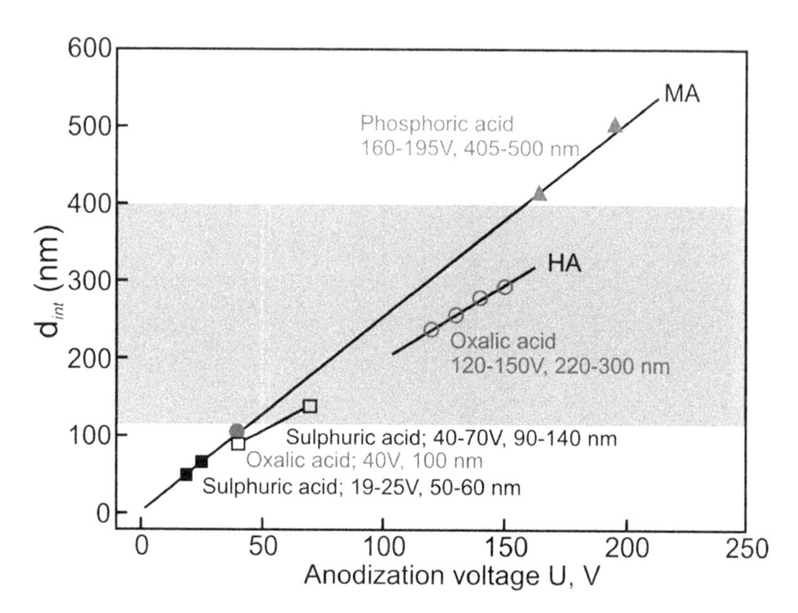

FIGURE 2.5 Dependence of the distance between pore centers (dint) on anodization voltage U.

Source: Reprinted with permission from Ref. [16]. Copyright 2006 Springer Nature.

2.4.2 Pore Diameter and Porosity of Anodic Aluminum Oxide Films

For application of anodic aluminum oxide as membranes, the pore diameter of oxide films is a very important parameter since it determines the predominant mechanism of gas diffusion through the membrane pores in case of gas separation, or the average size of separated particles—in case of filtration processes.

According to the model of oxide film ordering proposed in the paper,[31] the porosity is about 10% for hexagonally ordered porous structure of anodic aluminum oxide formed under "soft" anodization conditions. Then, combining formulas (2.3) and (2.5), it is possible to obtain the formula to calculate the average diameter of pores:

$$d_{\mathrm{p}} = \sqrt{\frac{2 \cdot \sqrt{3} \cdot \varepsilon}{\pi} k \cdot U_{\mathrm{volt}}}, \tag{2.6}$$

where ε is the porosity, k is the proportionality coefficient binding the anodization voltages and distance between pore centers.

However, this formula is applicable only for the conditions under which the hexagonal ordering of porous structure is observed. Besides, it is discussed in Ref. [16] that the porosity is 3.3–3.4% for the oxide films with hexagonal ordering of pores formed under "hard" anodization conditions, and, consequently, formula (2.5) is limited in application.

The additional method to control pore diameter and porosity of oxide film is the pore etching in solutions of different acids after anodization.[20–23] The microphotographs of anodic aluminum oxide films after etching in 5% (mass) phosphoric acid solution at 35°C during different time intervals are given in Figure 2.6. The pore diameter and porosity increase during etching, at the same time, the pore density remains constant.

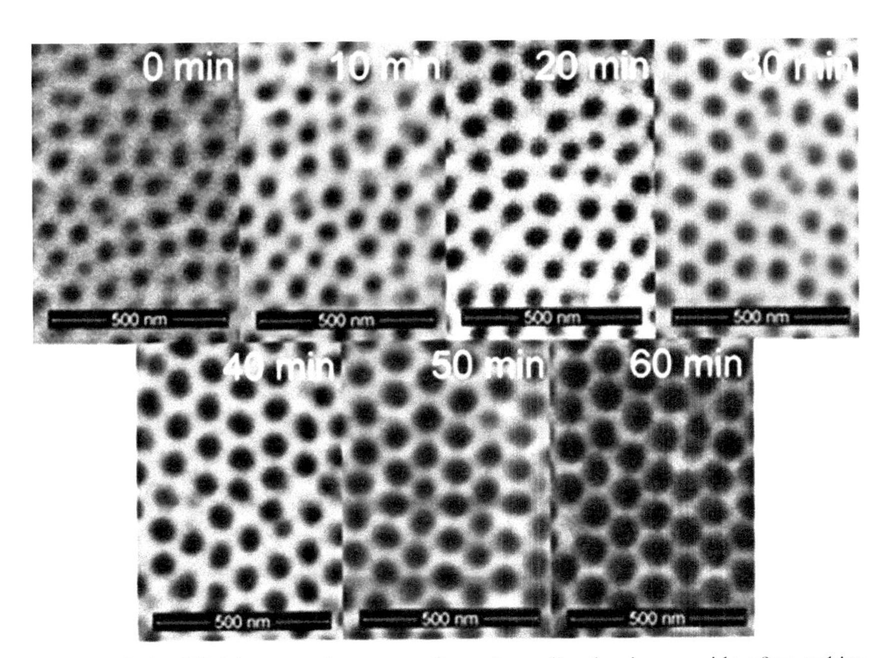

FIGURE 2.6 SEM images of upper surface of anodic aluminum oxide after etching during different time in 5% H_3PO_4 at room temperature (°C).

2.4.3 Thickness of Oxide Film

Although the process of aluminum anode oxidation is an electrochemical one, and, consequently, the mass of oxide formed (and thickness of oxide film) can be varied controlling the charge passing during anodization; in the majority of papers, the due regard is not paid to this approach and to control the thickness, the anodization processes is run during the fixed-time interval. The current output during galvanostatic anodization in the solutions of phosphoric and sulfuric acids at different current densities was studied in Refs. [24–26].

The reaction of anodic oxidation of aluminum can be written down as follows:

$$2\,Al + 3\,H_2O \rightarrow Al_2O_3 + 3\,H_2.$$

Then, according to Faraday's law, the mass of the oxide film formed equals:

$$m_t\left(Al_2O_3\right) = \chi \cdot \frac{M_{Al_2O_3}Q}{\left(6F_{const}\right)}, \tag{2.8}$$

where $m_t(Al_2O_3)$ is the oxide mass, χ is the current output, $M_{Al_2O_3}$ is the molar mass of Al_2O_3, Q is the total charge passing during anodization, F_{const} is the Faraday constant (96,485 C/mol)

It should be pointed out that during anodization a part of aluminum oxide formed dissolves with pore formation in the film structure, then:

$$m_p\left(Al_2O_3\right) = m_t\left(Al_2O_3\right) - m_d\left(Al_2O_3\right), \tag{2.9}$$

where $m_p\left(Al_2O_3\right)$ is the mass of the porous film, $m_d\left(Al_2O_3\right)$ is the mass of aluminum oxide dissolved. According to Faraday's law, the mass of metal aluminum $\left(m_{Al}\right)_\chi$ oxidized to oxide can be calculated as follows:

$$\left(m_{Al}\right)_\chi = \frac{\eta M_{Al}Q}{\left(3 \cdot F_{const}\right)}, \tag{2.10}$$

where M_{Al} is the molar mass of aluminum, $\left(m_{Al}\right)_\chi$ can be calculated taking into account mass balance during sample anodization using the following formula:

$$m_p(Al_2O_3) = (m_{Al})_\chi + m_t(Al_2O_3) - m_i(Al), \tag{2.11}$$

where $m_i(Al)$ and $m_t(Al_2O_3)$ are masses of samples before and after anodization, respectively.

Then, the current output can be calculated as follows:

$$\chi = \frac{(m_{Al})_\chi}{(m_{Al})_{\chi=100\%}}, \tag{2.12}$$

where $(m_{Al})_\chi$ is calculated using formula (2.11) with the values of $m_p(Al_2O_3)$, $m_t(Al_2O_3)$, and $m_i(Al)$ known from the experiment, and $(m_{Al})_{\chi=100\%}$ is found by formula (2.10).

According to the investigation results, the current density increases during galvanostatic anodization, regardless of electrolyte used, results in current output increase. Thus, during anodizing in 0.4 M solution of H_3PO_4 with current density of 5 mA/cm^2, the current output is 69%, with current density of 12 mA/cm^2—93%.[26]

The values of current output under 100% can be explained by the formation of ions, which do not participate in electrochemical reaction of aluminum oxidation taking place in the interface "metal/oxide" as a result of diffusion of O^{2-} ions through the barrier layer. At the same time, the reaction of water oxidation with the formation of gaseous oxygen can occur at the interface "oxide/electrolyte"; this reaction decreases the number of O^{2-} ions participating in the reaction of aluminum oxidation. The increase in anodization temperature leads to the increased ionic conductivity of barrier oxide layer; as a result, the current output goes up.[25] The investigations connecting the oxide film thickness with the passing charge were carried out in Ref. [27], where it is demonstrated that in case of potentiostatic anodization under the voltage of 160 V in 0.3 M H_3PO_4, the charge density of 1.924 C/cm^2 is required to form the oxide film 1 μm thick.

2.5 USING OF POROUS ALUMINUM OXIDE TO OBTAIN SEMICONDUCTORS NANOSTRUCTURES

The unique geometric parameters of the porous structure of porous aluminum oxide films, such as high-order degree and small variation of diameters of hexagonally distributed pores relative to the surface, arouse strong interest in the field of the synthesis of semiconductor nanoparticles and nanocomposites. Electrochemical methods are mainly used to obtain such objects, and porous aluminum oxide is used as a template with the pores filled with the required semiconductor. Chemical methods, in particular, electrochemical deposition and metal–organic chemical vapor deposition (MOCVD) are the most effective and frequently used methods of pore filling. Thus, CdSe,[28,29] CdS,[30,31] ZnO,[32] Ge,[33] CuS,[34] as well as ZnS[35–37] were successfully deposited.

The nanotubes of zinc sulfide were synthesized by MOCVD method in the work of Zhai et al.[36] (Fig. 2.7a,b). The powder of zinc bi-diethyl-dithiocarbamate $[Zn(S_2CNEt_2)_2]$ was evaporated in specially created reactor at 400°C to the colder, in comparison with the crucible temperature (approximately by 150°C), template of porous aluminum oxide. The authors demonstrated that the samples obtained during the radiation excitation with the wavelength of 375 nm have the luminescence with the wavelength of 510 nm (Fig. 2.7c).

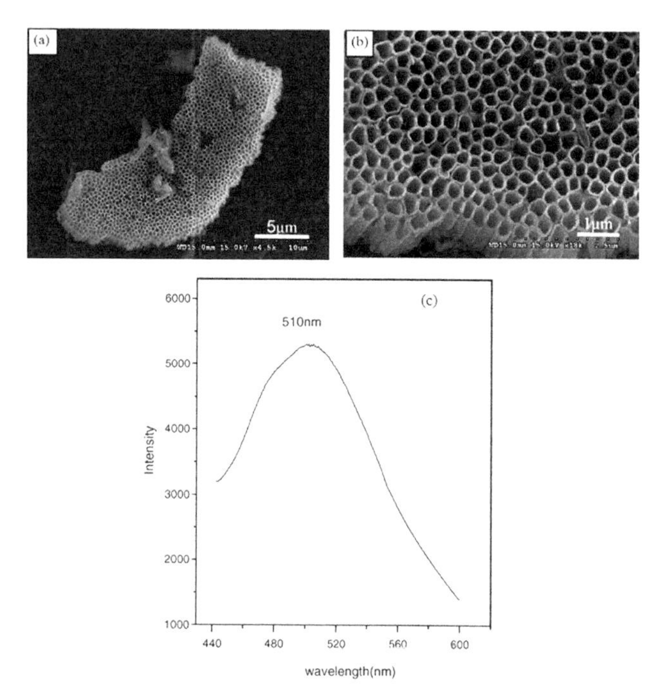

FIGURE 2.7 SEM patterns of ZnS nanotubes (a,b) and their luminescence spectrum under the excitation with the wavelength of 375 nm (c).
Source: Reprinted with permission from Ref. [36]. Copyright 2005 Elsevier B.V.

Xu[37] investigated the formation mechanism of ZnS nanowires by electrochemical deposition method. The solution of $ZnCl_2$ and elemental sulfur in dimethylsulfoxide was taken as electrolyte. The gold film playing the role of electrode was evaporated onto one of the sides of anodic aluminum oxide membrane before the deposition. The deposition reaction can be written down as follows:

$$8\ Zn^{2+} + S_8 + 16\ e = 8\ ZnS \quad (2.13)$$

The reaction was carried out at the current density of 6.62 mA/cm^2 and temperatures from 120 up to 130°C. After the deposition, the samples were washed in dimethylsulfoxide, and for the investigations by electron microscopy method, AAO membrane was etched in 0.5 M solution of NaOH.

The images of samples obtained by the methods of transmission (a) and raster (b) electron microscopy (TEM and REM, respectively) are given in Figure 2.8. The average diameter of nanowires of 40 nm, equaled to the matrix pore diameter, was determined with the help of TEM pattern. In scanning electron microscopy (SEM) pattern, it is seen that the nanowires are of the same length. The structural investigations by the method of X-ray diffraction (Fig. 2.8c) demonstrated that the nanowires have wurtzite-type hexagonal structure.

The authors also compared the above nanowires with the samples synthesized with the current density of 8.84 mA/cm^2. It was pointed out that the reaction rate significantly increases, but the X-ray phase analysis demonstrated the availability of phases of pure zinc and sulfur, together with ZnS, that is connected with the increased motion speed of Zn^{2+} ions, which are then combined into Zn nanoclusters, to the side of nanopores, and the availability of S phase is connected with the deposition of elemental sulfur into the pores directly from the solution.

As seen from the above examples of chemical deposition, this method has both advantages, for example, the possibility of complete filling of pores, and disadvantages, such as the availability of unreacted residues of chemical reactions in the samples. The method of deposition into AAO pores, which has been developed by our team for the last 10 years, is completely lacking the second disadvantage: thermal deposition under the conditions of high and ultrahigh vacuum.[38] With the help of this method, we obtained Ge,[39] ZnSe,[40,41] GaAs,[42] as well as ZnS[43] semiconductors nanostructures.

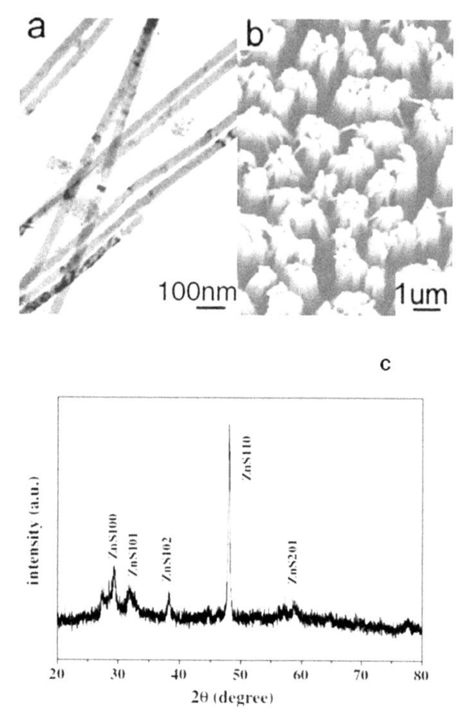

FIGURE 2.8 TEM (a) and SEM (b) patterns, as well as X-ray diffractogram (c) of zinc sulfide nanowires obtained by the deposition onto AAO membrane with the current density of 6.62 mA/cm^2.

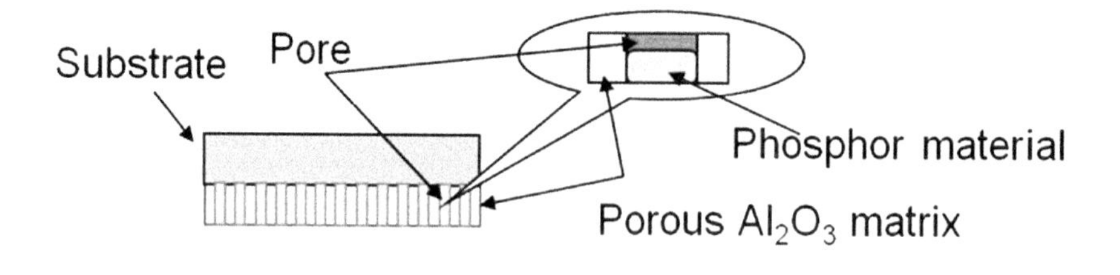

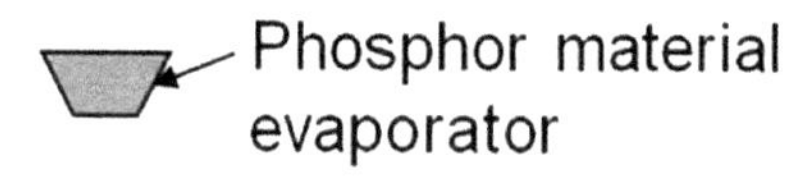

FIGURE 2.9 Scheme of thermal deposition of semiconductors onto the matrixes of porous AAO.

The scheme demonstrating the process of material deposition by the thermal method is given in Figure 2.9. The evaporation is carried out in vacuum chamber evacuated to high (not worse than 10^{-3} Pa) or ultrahigh (not worse than 10^{-7} Pa) vacuum. The powder of the evaporated material is put into the evaporator heated up to 1000°C or poured from vibrating bunker into evaporator heated up to 2000°C and then deposited onto the matrixes of porous AAO. More details of the processes running during deposition as well as more detailed description of the equipment applied will be given in Chapter 3.

SEM patterns of the obtained structures are given in Figure 2.10. It is found that the type of nanostructures can be described with the help of aspectual ratio A of the distance between pore centers d_{int} to their diameter d_{p}:

$$A = \frac{d_{\text{int}}}{d_{\text{p}}}. \tag{2.14}$$

In case of formation of rods in pore channels, A is in the range between 2.1 and 2.5. When A values are less than 2, it actually means that the wall thickness between neighboring pores is less than the diameter of these pores, the tubes are formed. Thus, the nanostructure formation can be conditionally split into two competing processes. The first—material adhesion on the pore walls, which results in its healing to the center. The second—film growth on the matrix surface, which results in closing the pores from the top.

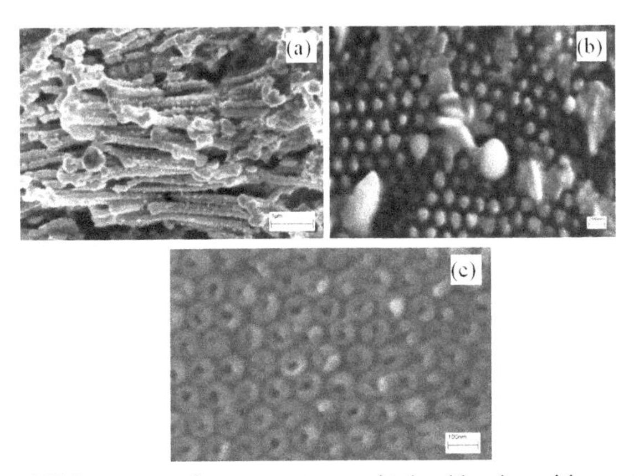

FIGURE 2.10 SEM patterns of nanostructures, obtained by deposition onto the matrixes of Al_2O_3 with pore diameter of 130 nm: (a) Ge, (b) ZnS, and (c) ZnSe.

KEYWORDS

- **anodic aluminum oxide**
- **electrochemical oxidation**
- **structural characteristics**
- **semiconductors**
- **electrochemical deposition**
- **thermal deposition**

REFERENCES

1. Bengough, G. D.; Stuart, J. M. Improved Process of Protecting Surfaces of Aluminium of Aluminium Alloys. *U.K. Patent 223,994*, 1923.
2. Jessensky, O.; Müller, F.; Gösele, U. Self-organized Formation of Hexagonal Pore Arrays in Anodic Alumina. *Appl. Phys. Lett.* **1998,** *72*, 1173–1175.
3. Masuda, H.; Fukuda, K. Ordered Metal Nanohole Arrays Made by a Two-Step Replication of Honey-Comb Structures of Anodic Alumina. *Science* **1995,** *268*, 1466–1468.
4. Masuda, H.; Yamada, H.; Satoh, M.; Asoh, H.; Nakao, M.; Tamamura, T. Highly Ordered Nanochannel-Array Architecture in Anodic Alumina. *Appl. Phys. Lett.* **1997,** *71*, 2770–2772.
5. Shingubara, S. Fabrication of Nanomaterials Using Porous Alumina Templates. *J. Nanopart. Res.* **2003,** *5*, 17–30.
6. Valeev, R. G.; Stashkova, V. V.; Chukavin, A. I.; Volkov, V. A.; Alalykin, A. S.; Syugaev, A. V.; Beltiukov, A. N.; Gilmutdinov, F. Z.; Kriventsov, V. V.; Mezentsev, N. A. Ni Nanostructures in Porous Anodic Alumina Matrices: Structure and Cathodic Properties in Hydrogen Release Reactions. *Phys. Proc.* **2016,** *84*, 407–414.
7. Conway, B. E.; Bockris, J. O. M.; White, R. E., Ed. *Modern Aspects of Electrochemistry*; Plenum Press: New York, London, 1989; Vol. 20, p 503.
8. Thompson, G. E. Porous Anodic Alumina: Fabrication, Characterization and Applications. *Thin Solid Films* **1997,** *297* (1–2), 192–201.
9. Zhu, X. F.; Li, D. D.; Song, Y.; Xiao, Y. H. The Study on Oxygen Bubbles of Anodic Alumina Based on High Purity Aluminum. *Mater. Lett.* **2005,** *59* (24–25), 3160–3163.
10. Takahashi, H.; Fujimoto, K.; Nagayama, M. Effect of pH on the Distribution of Anions in Anodic Oxide Films Formed on Aluminum in Phosphate Solutions. *J. Electrochem. Soc.* **1988,** *135* (6), 1349–1353.
11. Li, A. P.; Muller, F.; Birner, A.; Nielsch, K.; Gosele, U. Hexagonal Pore Arrays with a 50–420 nm Interpore Distance Formed by Self-organization in Anodic Alumina. *J. Appl. Phys.* **1998,** *84* (11), 6023–6026.
12. Chu, S. Z.; Wada, K.; Inoue, S.; Todoroki, S. Synthesis and Characterization of Titania Nanostructures on Glass by Al Anodization and Sol–Gel Process. *Chem. Mater.* **2002,** *14* (1), 266–272.

13. Young, L. *Anodic Oxide Films*; Academic Press: New York, 1961; p 377.
14. Poinern, G. E. J.; Ali, N.; Fawcett, D. Progress in Nano-engineered Anodic Aluminum Oxide Membrane Development. *Materials* **2011,** *4* (3), 487–526.
15. Diggle, J. W.; Downie, T. C.; Coulding, C. W. Anodic Oxide Films on Aluminium. *Chem. Rev.* **1969,** *69*, 365–405.
16. Lee, W.; Ji, R.; Gosele, U.; Nielsch, K. Fast Fabrication of Long-Range Ordered Porous Alumina Membranes by Hard Anodization. *Nat. Mater.* **2006,** *5* (9), 741–747.
17. Vanderlinden, B.; Terryn, H.; Vereecken, J. Investigation of Anodic Aluminum-Oxide Layers by Electrochemical Impedance Spectroscopy. *J. Appl. Electrochem.* **1990,** *20* (5), 798–803.
18. Osulliva, J. P.; Wood, G. C. Morphology and Mechanism of Formation of Porous Anodic Films on Aluminium. *Proc. R. Soc. Lond., Ser. A—Math. Phys. Sci.* **1970,** *317* (1531), 511–520.
19. Nielsch, K.; Choi, J.; Schwirn, K.; Wehrspohn, R. B.; Gosele, U. Self-ordering Regimes of Porous Alumina: The 10% Porosity Rule. *Nano Lett.* **2002,** *2* (7), 677–680.
20. Ersching, K.; Dorico, E.; da Silva, R. C.; Zoldan, V. C.; Isoppo, E. A.; Viegas, A. D. C.; Pasa, A. A. Surface and Interface Characterization of Nanoporous Alumina Templates Produced in Oxalic Acid and Submitted to Etching Procedures. *Mater. Chem. Phys.* **2012,** *137* (1), 140–146.
21. Zaraska, L.; Sulka, G. D.; Jaskula, M. Anodic Alumina Membranes with Defined Pore Diameters and Thicknesses Obtained by Adjusting the Anodizing Duration and Pore Opening/Widening Time. *J. Solid State Electrochem.* **2011,** *15* (11–12), 2427–2436.
22. Losic, D.; Losic, D. Preparation of Porous Anodic Alumina with Periodically Perforated Pores. *Langmuir* **2009,** *25* (10), 5426–5431.
23. Inada, T.; Uno, N.; Kato, T.; Iwamoto, Y. Meso-porous Alumina Capillary Tube as a Support for High-Temperature Gas Separation Membranes by Novel Pulse Sequential Anodic Oxidation Technique. *J. Mater. Res.* **2005,** *20* (1), 114–120.
24. Zhou, F.; Al-Zenati, A. K. M.; Baron-Wiechec, A.; Curioni, M.; Garcia-Vergara, S. J.; Habazaki, H.; Skeldon, P.; Thompson, G. E. Volume Expansion Factor and Growth Efficiency of Anodic Alumina Formed in Sulphuric Acid. *J. Electrochem. Soc.* **2011,** *158* (6), C202–C214.
25. Shawaqfeh, A. T.; Baltus, R. E. Growth Kinetics and Morphology of Porous Anodic Alumina Films Formed Using Phosphoric Acid. *J. Electrochem. Soc.* **1998,** *145* (8), 2699–2706.
26. Garcia-Vergara, S. J.; Habazaki, H.; Skeldon, P.; Thompson, G. E. Formation of Porous Anodic Alumina at High Current Efficiency. *Nanotechnology* **2007,** *8*, 416505 (8 pp).
27. Santos, A.; Formentin, P.; Pallares, J.; Ferre-Borrull, J.; Marsal, L. F. Structural Engineering of Nanoporous Anodic Alumina Funnels with High Aspect Ratio. *J. Electroanal. Chem.* **2011,** *655* (1), 73–78.
28. Gavrilov, S. A.; Kravtchenko, D. A.; Belogorokhov, A. I.; Zhukov, E. A.; Belogorokhova, L. I. Features of Luminescent Semiconductor Nanowire Array Formation by Electrodeposition into Porous Alumina. In *Physics, Chemistry and Application of Nanostructures: Reviews and Short Notes to Nanomeeting*—2001; World Scientific Publ. Co Pte. Ltd., Singapore, 2001; pp 317–320.

29. Laatar, F.; Hassen, M.; Amri, C.; Laatar, Fe.; Smida, A.; Ezzaouia, H. Fabrication of CdSe Nanocrystals Using Porous Anodic Alumina and Their Optical Properties. *J Luminesc.* **2016,** *178*, 13–21.
30. Zhao, Y.; Yang, X.-C.; Huang, W.-H.; Zou, X.; Lu, Z.-G. Synthesis and Optical Properties of CdS Nanowires by a Simple Chemical Deposition. *J. Mater. Sci.* **2010,** *45*, 1803–1808.
31. Sun, H.; Li, X.; Chen, Y.; Li, W.; Li, F.; Liu, B.; Zhang, X. The Control of the Growth Orientations of Electrodeposited Single-Crystal Nanowire Arrays: A Case Study for Hexagonal CdS. *Nanotechnology* **2008,** *19*, 225601 (8 pp.).
32. Hu, D.; Fan, D.; Zhu, Y.; Wang, W.; Pan, F.; Wu, P.; Fan, J. Synthesis of Location-Dependent Phosphorus-Doped ZnO Nanostructures on the Porous Alumina Membranes. *Phys. Stat. Sol. A* **2014,** *211* (4), 856–861.
33. Li, X.; Meng, G.; Qin, S.; Xu, Q.; Chu, Z.; Zhu, X.; Kong, M.; Li, A.-P. Nanochannel-Directed Growth of Multi-Segment Nanowire Heterojunctions of Metallic $Au_{1-x}Ge_x$ and Semiconducting Ge. *Nanotechnology* **2012,** *6* (1) 831–836.
34. Liang, C.; Terabe, K.; Hasegawa, T.; Aono, M. Template Synthesis of M/M_2S (M = Ag, Cu) Hetero-Nanowires by Electrochemical Technique. *Solid State Ionics* **2006,** *177*, 2527–2531.
35. Sun, H.; Yu, Y.; Li, X.; Li, W.; Li, F.; Liu, B.; Zhang, X. Controllable Growth of Electrodeposited Single-Crystal Nanowire Arrays: The Examples of Metal Ni and Semiconductor ZnS. *J. Cryst. Growth* **2007,** *307*, 472–476.
36. Zhai, T.; Gu, Z.; Ma, Y.; Yang, W.; Zhao, L.; Yao, J. Synthesis of Ordered ZnS Nanotubes by MOCVD-Template Method. *Mater. Chem. Phys.* **2006,** *100*, 281–284.
37. Xu, X.-J.; Feil, G.-T.; Yu, W.-H.; Wang, X.-W.; Chen, L.; Zhang, L.-D. Preparation and Formation Mechanism of ZnS Semiconductor Nanowires Made by the Electrochemical Deposition Method. *Nanotechnology* **2006,** *17*, 426–429.
38. Valeev, R.; Surnin, D.; Vetoshkin, V.; Beltukov, A.; Eliseev, A.; Napolskii, K.; Roslyakov, I.; Petukhov, D. A Method of Obtaining of Semiconductor Nanostructure. Russian Patent No. 2,460,166.
39. Valeev, R. G.; Surnin, D. V.; Beltyukov, A. N.; Vetoshkin, V. M.; Kriventsov, V. V.; Zubavichus, Ya. V.; Mezentsev, N. A.; Eliseev, A. A. Synthesis and Structural Study of the Ordered Germanium Nanorod Arrays. *J. Struct. Chem.* **2010,** *51*, S132–S136.
40. Valeev, R. G.; Deev, A. N.; Romanov, E. A.; Kriventsov, V. V.; Beltyukov, A. N.; Mezentsev, N. A.; Eliseev, A. A.; Napolskii, K. S. Synthesis and Structure Study of Ordered Arrays of ZnSe Nanodots. *Surf. Investig. X-Ray, Synchrotr. Neutron Tech.* **2010,** *4* (4), 645–648.
41. Valeev, R.; Romanov, E.; Deev, A.; Beltukov, A.; Napolski, K.; Eliseev, A.; Krylov, P.; Mezentsev, N.; Kriventsov, V. Synthesis of ZnSe Semiconductor Nanodots Arrays by Templated PVD. *Phys. Stat. Sol. C* **2010,** *7*, 1539–1541.
42. Valeev, R. G.; Kobziev, V. F.; Kriventsov, V. V.; Mezentsev, N. A. Synthesis and Structure Investigations of Nanostructures Massive of GaAs. *Bull. Russian Acad. Sci. Phys.* **2013,** *77* (9), 1157–1160.
43. Valeev, R.; Romanov, E.; Beltukov, A.; Mukhgalin, V.; Roslyakov, I.; Eliseev, A. Structure and Luminescence Characteristics of ZnS Nanodot Array in Porous Anodic Aluminum Oxide. *Phys. Stat. Sol. C* **2012,** *9*, 1462–1465.

CHAPTER 3

Vacuum-Thermal Deposition of Semiconductors

ABSTRACT

This chapter contains the description of the methods of the vacuum-thermal deposition of semiconductors. Processes of evaporation and deposition are described. Experimental approaches and equipment for vacuum-thermal deposition and evaporators are described in detail in the chapter.

3.1 PROCESSES OF EVAPORATION AND DEPOSITION

Thermal evaporation of the powder or target of the substance under high or ultrahigh vacuum is the most technological and simplest method of material obtaining. It attracts the attention of engineers and researchers by its simplicity, possibility to use the templates of different materials, as well as with the availability of template cooling and heating system, the possibility to obtain films in different structural states depending on deposition conditions.

If the vacuum chamber has rather large volume, it is possible to synthesize samples characterized by different amount of the material deposited. The application of this method for the synthesis of thin films of electrophosphors requires detailed investigation of conditions of deposition and their influence on the structure and properties. The method fundamentals and processes of substance evaporation and condensation on templates, as well as brief description of the equipment required are presented in the chapter.

The process of film formation by the vacuum method comprises four main stages: evaporation of initial materials, nucleation, surface migration of islands, and Ostwald ripening.[1]

The kinetic theory of gases allows interpreting the evaporation process. The evaporation theory comprises the elements of reaction kinetics, thermodynamics, and theory of solid. Kossel[2,3] and Stranski[4] proposed the model of crystal surface to describe evaporation of crystalline solids, mainly semiconductors (Fig. 3.1).

The crystal surface can be represented as the aggregation of atoms fixed in the crystal lattice site (*K*), in the surface plane (*S*) and placed on the free surface (*A*, *L*). The number of neighbors of atoms *S* and *L'* exceed the average value; consequently, it is necessary to supply energy exceeding the lattice energy for their evaporation.

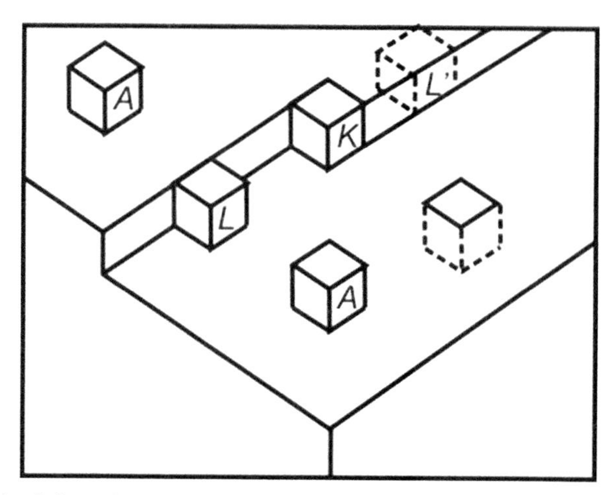

FIGURE 3.1 Schematic representation of crystal surface and atom positions with different number of atoms in the first and second coordination sphere. *S*—Surface atom inside the completely formed crystalline surface; *L'*—atom inside the monoatomic step; *K*—atom inside the crystal lattice site; *L*—atom on the monoatomic step; and *A*—single atoms adsorbed on the surface.

Although atoms *A* and *L* have less neighbors and can be removed from the surface with less energy consumption. The binding energy of atom *K* in the lattice crystal site equals the lattice energy. In this regard, it is supposed that a crystal can be constructed by successive addition of crystal site atoms, and the removal and addition of such atoms presents microscopically reversing events, which control the equilibrium of evaporation and growth of crystal surface. The probability of atom direct transition from the crystal lattice site to vapor calculated from the binding energy is very low to explain the observed evaporation rates. Therefore, Vollmer[5] developed the theory of evaporation process and demonstrated that the atom transition from the crystal lattice site (K) to the location on the edge (A, L) followed by the diffusion along the crystal surface precedes evaporation (Fig. 3.1). For the atom desorption to occur, it is necessary for the atom diffusing along the surface to obtain the necessary energy of translational motion.

The films are grown from the vapor phase formed by the atoms and molecules of components in closed evacuated containers or vacuum chambers. Condensation rate is the factor influencing the structural properties of the film formed. Condensation rate mainly depends on the evaporation rate of the initial material from the source surface. The evaporation rate is found by Hertz–Knudsen equation[6]:

$$\frac{\mathrm{d}N_\mathrm{e}}{S_\mathrm{e}\,\mathrm{d}t} = \alpha_\mathrm{v}\left(2\pi m_\mathrm{p} k_\mathrm{B} T_\mathrm{e}\right)^{-1/2}\left(P^* - P_\mathrm{h}\right), \tag{3.1}$$

where $\mathrm{d}N_\mathrm{e}/\mathrm{d}t$ is the evaporation rate, S_e is the evaporation area, a_n is the evaporation coefficient, m_p is the mass of atom or molecule, T_e is the evaporator temperature, P^* is the equilibrium pressure, and P_h is the hydrostatic pressure on the surface.

The mechanism of a new phase formation greatly influences the further growth of films, their structure, optic, and electric properties.

The formation of embryos (disperse particles) of a new phase is the inherent stage of any phase transition of first kind. Three main mechanisms of thin film growth are distinguished: Frank–van der Merve growth mode, Vollmer–Weber growth mode and Stranski–Krastanov growth mode.[7]

Frank–van der Merve growth mode occurs in case of strong adhesion, when the bond energy between the atoms of crystal-template and film is more than the energy between the film atoms. A new phase grows through the formation of monolayers. In this case, the formation of one or several layers of condensate on the template is advantageous thermodynamically, even if the vapor is saturated. The work of forming embryos on the template is insignificant. The mechanism of complete wetting is experimentally observed for isostructured pairs with very similar lattice parameters, the value of crystallographic discrepancy is under 1%. With the film growth by Frank–van der Merve, interatomic distances in atomic meshes of condensate and template, parallel to the interface, appear to be the same. The condensate film is either stretched or compressed in the template plane. Accordingly, the film is either compressed or stretched in directions perpendicular to this plane. If we switch off from this deformation, the condensate atoms are placed into the template lattice overbuilding and forming the so-called pseudo-amorphous layer. The epitaxial ratios in this case are obvious and come down to the coincidence of all similar crystallographic planes and directions of the template and condensate.

The island mode (Vollmer–Weber) is characteristic for weak adhesion of crystal on the template. The work of forming embryos on the template is greater than in the previous case and can even approach the work of embryo formation during the homogeneous nucleation. The role of irregularities on the template decreasing the embryo-formation barrier is growth dramatically. A new phase is formed by the origination of 3D embryos. Weak adhesion results in less prominent bond of crystal orientations—condensate with template. The island growth takes place, if the following condition is fulfilled:

$$\sigma_{\mathrm{s}} < \sigma_{\mathrm{d}} < \sigma_{\mathrm{s-d}} - \mathrm{const} \times k_{\mathrm{B}} T_{\mathrm{e}} \ln\left(\xi_{\mathrm{ov}} + 1\right), \tag{3.2}$$

where σ_{s} is the free energy of template surface unit, σ_{d} is the free energy of adsorbent surface unit, $\sigma_{\mathrm{s-d}}$ is the free energy of interface "template-adsorbent" surface unit, x_{ov} is the oversaturation.

Otherwise, the layer-by-layer mode takes place. In island mode, small embryos are formed directly on the template surface and then they grow, transforming into big islands of condensed phase.[8] When blending, these islands form a uniform film after filling the channels between them.[9]

In intermediary mode, Stranski–Krastanov mode, the first layer completely covers the template surface, and 3D islands of film are growing on it. The main reason—change in mechanisms. The lattice parameter cannot remain unchanged when the next layer is filled. Its change results in high energy increase on the interface of surface "adsorbent-intermediary layer," which provides the fulfillment of island mode criterion.

Multiple experimental investigations demonstrate that at the initial stages of crystalline film condensation on foreign crystalline templates the embryos of a new phase can relatively quickly move along the template surface.[10–12] This process is an important link in the film structure formation. Such movements take place under the action of different external forces: encounter with fast particles of the flow, gradient of temperatures, and force fields of different origin.

At the moment, three main models are used to describe the migration mechanism of islands.[7] In the first model, the diffusion of particles takes place only along the island surface, and the particles adjoining the interface surface remain motionless relative to it. In the second model, the island is considered as sliding along the template surface. In the models of the first type, this is connected with low values self-diffusion coefficients of adatoms, which cannot provide relatively fast movement of the island. In the models of the second type, this is connected with high sliding friction forces between the island and template (i.e., high energy of sliding activation). These two mechanisms result in relatively low values of migration rate of islands and their diffusion coefficient. The third model of the movement of directly growing islands along the foreign template was proposed in Ref. [13]. As it is known, under certain conditions, the discrepancy dislocations occur on the interface "island-template." In this case, when Burgers vector lies in the sliding plane, the island movement can be provided by the motion of these dislocations (solitons).[13] Such mechanism looks more preferable than ordinary sliding due to the fact that during the dislocation motion nearly all atoms of the island remain motionless relative to the template at any time, and only a very small group of atoms moves. The dislocation passing from one part of the island to another one amounts to the island transition to one lattice spacing of the template.

The stage of Oswald ripening (OR) is responsible for the film structural formation, and, consequently, for the properties of the films grown.[13–17]

The stage of OR is the late stage of phase transition. It starts only with rather weak sources of deposited atoms, when oversaturation is low and tends to zero. In this case, new islands are not formed. The physics of OR is as follows. At the late evolution stage of the island assembly, a peculiar interaction between them occurs. This interaction is carried out through the generalized self-consistent diffusive field, which can be formed on the template by adatoms with the concentration ρ_a, atoms in vapor stage with the density ρ_v, or, if there are linear defects on the template, by adatoms adsorbed near steps with the concentration ρ_l. This field depends on the function of island distribution by size $f(R,t)$ and is in equilibrium with the islands with the critical size R_c. In case of vanishing sources of substance, the assembly of dispersed particles (DPs) takes the form of δ-function with $R - R_{c0}$. With nonvanishing sources of substance, the system tends to δ-function with $R - R_{cp}$. Islands with the sizes $R < R_c$ are dissolved in the diffusive field, as the equilibrium concentration of atoms ρ_R near them is greater than the average concentration of the field $\rho_R > \rho_a$, $\rho_R > \rho_l$. Islands with the sizes $R > R_c$ grow, as for them $\rho_R < \rho_a$, $\rho_R < \rho_l$. The critical size itself R_c is constantly increasing, since the islands absorb the substance from the template, thus decreasing oversaturation.

The authors[18] established equations for the growth kinetics of DPs at different mechanisms of mass transfer. The following two cases can be distinguished in case of DP growth during the diffusion of components in volumetric vapor phase:

(1) If the growth rate is limited by diffusion of components in vapor phase, the equation for the rate of change of DPs can be written down as follows:

$$\frac{dR}{dt} = \frac{2\sigma D_G V^2 \psi_1(\Theta)\alpha(\Theta)}{k_B T R^2}\left(\frac{R}{R_c} - 1\right), \tag{3.3}$$

where σ is the surface tension, D_G is the diffusion coefficient in vapor phase, V is the volume per one atom in solid phase, $\psi_1(\theta)$, $\alpha(\theta)$ is the parameters taking into account the shape of DP, k_B is the Boltzmann constant, T is the temperature, R is the radius of DP, and R_c is the critical radius of DP.

(2) If the growth rate and incorporation of atoms into the crystalline lattice are limited, the equation is as follows:

$$\frac{dR}{dt} = \frac{2\sigma \beta_G V^2 \psi_l(\Theta)\alpha(\Theta)}{k_B T R^2}\left(\frac{R}{R_c} - 1\right), \tag{3.4}$$

where β_G is the specific boundary on DP characterizing the formation rate of chemical bonds.

When DP grows by the mechanism "evaporation-condensation," the formula for DP growth rate is as follows:

$$\frac{dR}{dt} = \left(\frac{P^2\sqrt{2\pi m k_B T}}{P^*}\right)\frac{2\sigma V^2 \psi_1(\Theta)\alpha(\Theta)}{k_B T R}\left(\frac{R}{R_c} - 1\right), \tag{3.5}$$

where P is the component pressure in vapor around DP, and P^* is the component equilibrium pressure in vapor.

If DP grows during the substance diffusion along the template surface, the equation of DP radius change rate, in general, is as follows:

$$\frac{\mathrm{d}R}{\mathrm{d}t}=\frac{VJ_{\mathrm{SR}}}{R}\psi(\Theta), \tag{3.6}$$

where V is the atom volume, J_{SR} is the adatom flow, $\psi(\theta)$ is the multiplier taking into account the fact that the particle has the shape of spherical segment.

If DP grows by 1D diffusion of the substance along the steps, the formula of DP size change rate is as follows:

$$\frac{\mathrm{d}R}{\mathrm{d}t}=\frac{M_{\mathrm{st}}VJ_{\mathrm{SR}}}{4\pi R^{2}}\psi_{1}(\Theta), \tag{3.7}$$

where M_{st} is the number of steps crossing DP and J_{SR} is the substance flow along the steps.

Usually, DP grows simultaneously by all mass-transfer mechanisms described (evaporation–condensation, diffusion along the surface, diffusion along the steps) that complicates the practical application of the equations. However, when creating certain conditions, such as the temperature mode, the power of the substance sources, the source vanishing degree, one of these mechanisms can become predominant.

3.2 EXPERIMENTAL APPROACHES AND EQUIPMENT

3.2.1 Experimental Setups for Vacuum-Thermal Deposition

The typical setup for vacuum-thermal deposition represents a vacuum chamber equipped with evacuation system for high of ultrahigh vacuum. The typical scheme of setup is given in Figure 3.2.

The ion pump **NM** with ultimate vacuum under 10^{-7} Pa was used as the primary pump to reach the operating vacuum in the chamber not worse than 10^{-6} Pa. This pump **NM** together with the deposition chamber **CT** and mass-spectrometer **S** represent the heated up ultrahigh vacuum part of the whole system. The turbomolecular pump **NR** is applied for evacuation from the vacuum chamber during its degassing. Such pump cannot be used as the primary one due to insufficient compression toward light active gases (H_2, H^-, H_2O, O^-, N^-, OH^-, etc.). The prevacuum before the starting pressure of the turbomolecular pump (P = 1 Pa) is created by the rotary-vane pump **NL**. At the first evacuation stage, the vacuum control is executed by the thermocouple gauge **PT1** (up to 10^{-2} Pa). The sensor allows determining the operation efficiency of the trap **B** cooled down by running water or liquid nitrogen, as well as the ultimate vacuum in the turbomolecular pump with the closed high-vacuum valve **VH**.

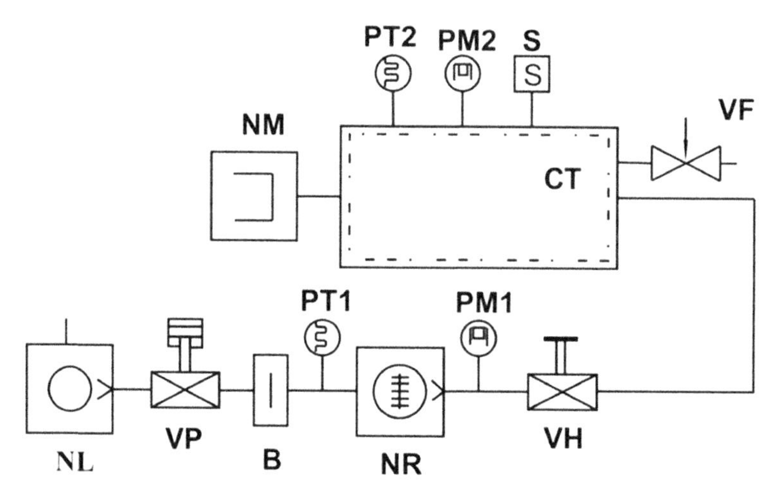

FIGURE 3.2 Scheme of vacuum part of the plant for vacuum-thermal deposition. **NL**—Rotary-vane pump; **VP**—pneumatic valve; **B**—nitrogen trap; **PT1**, **PT2**—thermocouple gauges; **CT**—heated up vacuum chamber; **PM1**, **PM2**—Penning gauges; **NR**—turbomolecular pump TMN-500; **NM**—ion pump; **VF**—overlap; **S**—mass-spectrometer; and **VH**—manual high-vacuum valve.

The Penning gauge **PM1** (to 10^{-7} Pa) allows controlling the residual pressure of the turbomolecular pump. The valve **VP** is intended to close the access to the turbomolecular pump at the initial start moment of the rotary-vane pump.

During the operation of the ion pump (starting from 10^{-2} Pa) the vacuum chamber is separated from the turbomolecular pump by the manual valve **VH**. The valve **VF** is intended for the atmosphere inflow to all nodes of the vacuum chamber.

Evaporators, templates, template holders, template heaters, and gates are the main elements of the deposition chamber (Fig. 3.3).

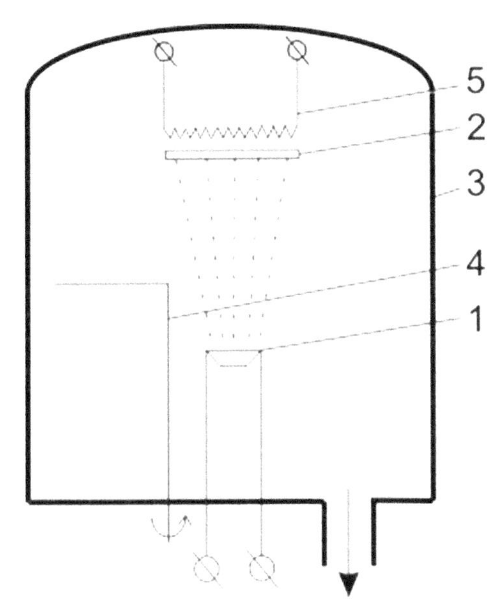

FIGURE 3.3 Scheme of vacuum-thermal deposition chamber. 1—Evaporator; 2—template; 3—chamber; 4—gate; and 5—template heater.

3.2.2 Evaporators

The following requirements are specified for the evaporator materials:

(1) there must be no chemical reactions between the evaporator material and substance evaporated;
(2) there must be no highly volatile alloys of these materials formed, as otherwise, the films applied are contaminated and the evaporator destructs; and
(3) vapor pressure of the evaporator material at the evaporation temperature of the deposited substance must be negligible.

During vacuum thermal deposition the evaporated materials are heated directly or indirectly: during the direct heating—immediately with the current passed through it (resistance heating), induction way, or electron bombardment; during the indirect heating—with the heat transfer from the evaporator. Moreover, the evaporator heating can be also resistance, inductive, or electron bombarding.

Direct resistance heating is applied for the materials with the melting temperature above the evaporation temperature (evaporation by the sublimation method). Such material comprise: zinc sulfide and selenide, zinc, chrome, titanium, manganese, magnesium, and cadmium.

The main advantage of this method consists in the lack of heat contact between the evaporated material and evaporator elements, resulting in the improved purity of the deposited film. The main limitation is impossibility of dielectrics evaporation. Besides, it is impossible to evaporate the majority of metals by resistance direct heating.

Induction heating is characterized by high equipment cost (powerful high-frequency generators are required). The method of direct heating of the material evaporated by electron bombardment allows obtaining the energy flow with high power concentration (up to $5–10^8$ W/cm^2) that is necessary to evaporate heat-resistant materials and dielectrics.

Evaporators with indirect resistance heating are used to evaporate metals, semiconductors, and dielectrics. In case of small amount, the metal evaporated is used in the form of wire (Fig. 3.4a). When the current passes, the metal evaporated melts and wets the evaporator. Thin tungsten wire increases the surface wetted and thermal contact of the material evaporated with the evaporator. The band evaporator (Fig. 3.4b) produced from thin sheet of heat-resistant material with molded hemispheres is applicable for the evaporation of loose materials (metals and dielectrics).

If during deposition the ejection of macroscopic particles damaging the film deposited is observed, evaporators in the form of a boat (Fig. 3.4c) with one or two screens are applied, the holes in which are shifted checker-wise. The crucible evaporator demonstrated in Figure 3.4d is used to evaporate large quantities of loose dielectric materials.

Some materials (aluminum, in particular) during evaporation from metal evaporators (wire, band, and boat-type) with indirect heating form volatile alloys that results in decomposition of evaporators and contamination of films. In this case, evaporation from crucible evaporators is carried out (Fig. 3.4d–f). The crucible material is selected to lack chemical reactions with the material evaporated. It is necessary to take into account the evaporation temperature when selecting the crucible material. The graphite crucible evaporator given in Figure 3.4f represents the rod with milled out recess in the center into which the substance to be evaporated is put.

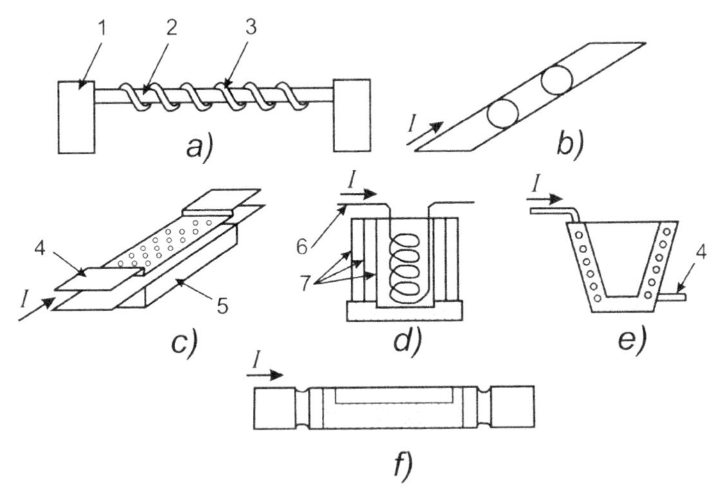

FIGURE 3.4 Evaporators with resistance (a) and indirect (b–f) heating. 1—Current leads, 2—tungsten heater, 3—tungsten wire of small diameter, 4—screen, and 5—boat, 6—wire heater, and 7—thermal screen.

More advanced evaporating cells of molecular beam epitaxy are applied in modern technological plants (Fig. 3.5).

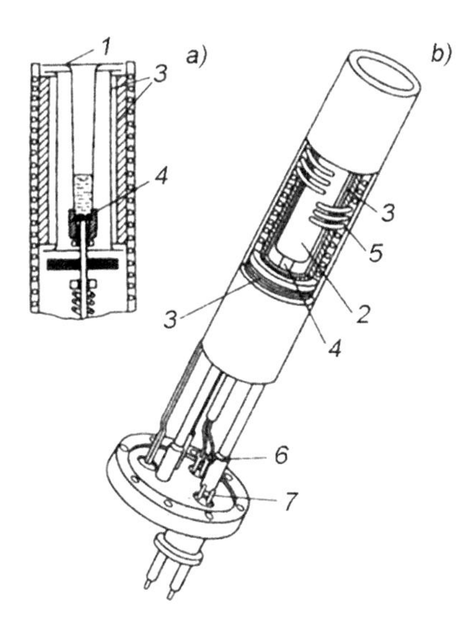

FIGURE 3.5 Heating part of the cell of molecular beam epitaxy (a) and external view in the assembled condition (b). 1—Gate, 2—crucible, 3—thermal screens, 4—thermocouple, 5—heater, 6—thermocouple current leads, and 7—heater current leads.

The feedback that allows adjusting the evaporator temperature to maintain the constant rate of the film growth is formed with the help of special electric circuits between the heater and thermocouple and sometimes also special devices to measure the thickness of the material deposited. The selection of evaporation mode of such cells allows achieving atom-layer growth of films with perfect crystalline structure.

KEYWORDS

- **thermal deposition**
- **thin films**
- **deposition processes**
- **experimental setup**
- **evaporators**

REFERENCES

1. Belyi, A.; Karpenko, G.; Myshkin, N. *Structure and Methods of Formation of Wear-Resistant Surface Layers*; Mashinostroenie, Moscow, 1991; p 208 (in Russian).
2. Kossel, W. Zur Theorie des Kristallwachstums. *Nach. Ges. Wiss. Göttingen, Math.-Phys.* **1927,** *K1*, 135–143.
3. Kossel, W. Existenzbereiche von Aufbau- und Abbauvorgangen auf der Kristallkugel. *Ann. Phys.* **1938,** *33*, 651.
4. Stranski, I. N. Zur Theorie der Kristallwachstums. *Z. Phys. Chem.* **1928,** *136*, 259–278; **1931,** *11B*, 421.
5. Ed. By Pelzer, H.; Volmer, M., Ed. *Kinetik der Phasenbildung*; Teodor Steinkopff Verlag, Dresden, 1939; p 220.
6. Messel, L.; Glang, R., Ed. *Handbook of Thin Film Technology*; McGraw Hill Hook Company, New York, 1970; p 800.
7. Kukushkin, S. A.; Osipov, A. V. Thin-film Condensation Processes. *Phys. Usp.* **1998,** *41*, 983–1014.
8. Venables, A.; Spiller, G. D. T.; Hanbucken, M. Nucleation and Growth of Thin Films. *Rep. Prog. Phys.* **1984,** *47* (4), 399–459.
9. Pashley, D. W. The Nucleation, Growth, Structure, and Epitaxy of Thin Surface Films. *Adv. Phys.* **1965,** *14*, 327–416.
10. Basset, G. A. *Condensation and Evaporation of Solids*; Gordon and Breach, New York, 1964; 599 p.
11. Kern, R.; Masson, A.; Metios, J. J. Migration Browhienne de Crisstallites Sur. Une Surface et Relation Avee Depitaxie. *Surf. Sci.* **1971,** *27* (3), 483–498.

12. Zanghi, J. C.; Metios, J. J.; Kern, R. Elastic Interaction between Small Nuclei. *Surf. Sci.* **1975,** *52* (3), 556–568.
13. Kukushkin, S. A.; Osipov, A. V. Soliton Model of Island Migration in Thin Films. *Surf. Sci.* **1995,** *329* (1–2), 135–140.
14. Kalinkin, I.; Aleskovsky, V.; Simashkevich, A. A. *Epitaxial Films of A^2B^6 Compounds*; Zhdanov's LSU, Leningrad, 1978; p 311 (in Russian).
15. Ivlev, V.; Trusov, A.; Kholmyansky, V. *Structural Transformations in Thin Films*; Metallurgia, Moscow, 1988; p 325 (in Russian).
16. Palatnik, L.; Papirov, I. *Epitaxial Films*; Nauka, Moscow, 1971; p 480 (in Russian).
17. Palatnik, L.; Fuks, M.; Kosevich, V. *The Mechanism of Formation and Substructure of Condensed Films*; Nauka, Moscow, 1972; p 319 (in Russian).
18. Kukushkin, S.; Slezov, V. *Dispersed Systems on the Surface of Solid States: Mechanism of Thin Films Formation (An Evolutional Approach)*; Nauka, Saint-Petersburg, 1996; p 304 (in Russian).

CHAPTER 4

Methods of Investigating Structure and Chemical Composition

ABSTRACT

This chapter describes methods of investigating structure and chemical composition. In this chapter, X-ray diffraction, extended X-ray absorption fine structure (EXAFS) spectroscopy, and application of Fourier fitting to extract the structural information from EXAFS spectra and X-ray electron spectroscopy are discussed.

4.1 X-RAY DIFFRACTION

Electrophosphors on the basis of A^2B^6 compounds typically represent powder materials or polycrystalline films. The crystalline structure of such materials can be investigated by X-ray, electron, and neutron diffraction methods. From methodological point, X-ray structural analysis is the simplest method providing rather complete information. In most cases, X-ray structural investigations are carried out on powder diffractometers in Bragg–Brentano focusing. X-ray tube and detector slit are located on the circumference with the plane sample in its center (Fig. 4.1).

Depending on goniometer type, the sample turns by angle θ and detector—by angle 2θ (θ–2θ goniometer) during scanning, or detector and X-ray tube turn by the same angle with the fixed sample (θ–θ goniometer). In both cases, the sample can additionally rotate around the normal line to the surface to diminish the error of determining the reflection intensity from coarse-grained samples.

The investigation of phosphors crystalline structure is critical for the formation of light-emitting devices. Thus, for instance, zinc sulfide can be formed in two different phases—cubic (sphalerite) and hexagonal (wurtzite). At the same time, the cubic phase has more effective luminescence, and the admixture of hexagonal phase can result in decreasing the electroluminescence brightness.[2]

In some papers, it was demonstrated that electrophosphors with defects in crystalline lattice caused by the transition from wurtzite to sphalerite possess the highest intensity.[3–6] Thus, the characterization of crystalline structure is the most important task in electrophosphors formation technologies.

The typical diffraction pattern of ZnS powder is given in Figure 4.2. The cubic phase is the basis, however, the peaks of small wurtzite intensity are also observed. The emergence of hexagonal phase is connected with the availability of packing defects, whose appearance energy for this material is relatively low. This is especially significant for thin films, during the growth of which, as a result of packing errors, twinning planes forming hexagonal interlayers emerge.[7]

In this regard, the significant part played is the selection of substrate, on which the film is formed. The discrepancy of parameters of substrate lattices and film being formed results in microstresses.

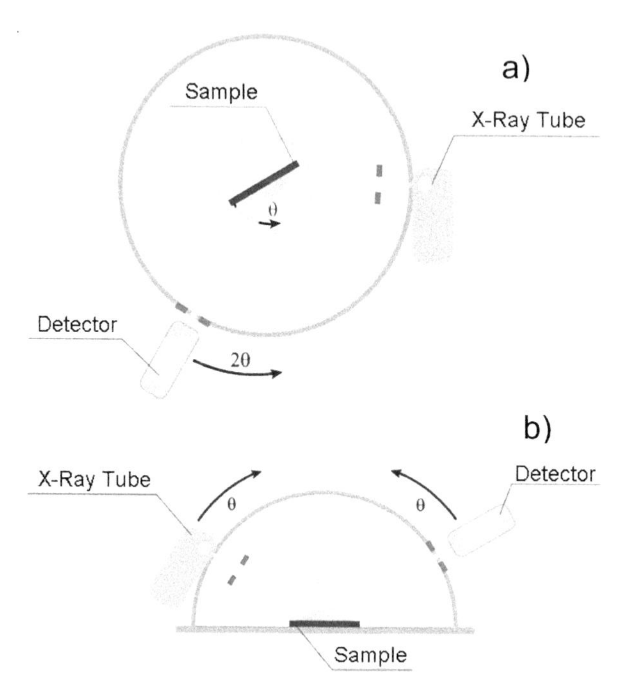

FIGURE 4.1 Operation scheme of θ–2θ goniometer (a) and θ–θ goniometer (b).

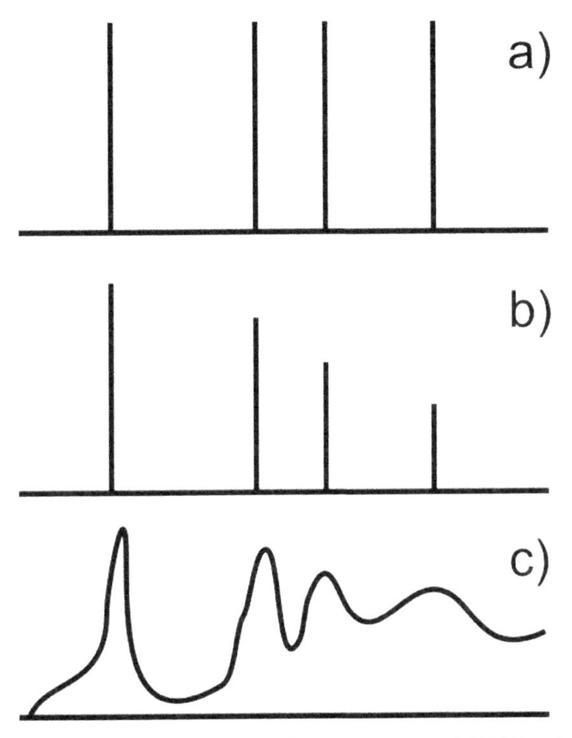

FIGURE 4.2 An influence of defects on diffraction pattern. (a) Diffraction pattern from ideal crystal, (b) effect from I class of defects, and (c) effect from II class of defects.

In papers [6, 8], zinc sulfide films were deposited by the method of flash evaporation onto Si, SiO_2, and NaCl substrates. In case of silicon and quartz, the films had only the cubic modification, but the films based on mineral salt had also the hexagonal phase, which disappeared during annealing. When the deposition conditions were changing, the hexagonal phase appeared in the films and substrates of Si, SiO_2, but the annealing did not lead to the structural changes. Such behavior is connected with the different sign of mechanical stresses in the system ZnS/NaCl, as compared with the systems ZnS/Si and ZnS/SiO_2. The microstresses are characterized by the uniform compression or extension of the lattice in the area of elastic deformations that result in the uniform change in interplanar distances by the value Δd_{hkl}. In X-ray diffraction patterns, this results in the displacement of maximums by angle $\Delta\theta_{hkl}$. Having differentiated Wulff–Bragg equation ($2d \sin \theta = n\lambda$), we can obtain the expression for the relative deformation by the sign of which the deformation type is determined—compression or extension:

$$\varepsilon = \frac{\Delta d_{hkl}}{d_0} = -\Delta\theta_{hkl} ctg\theta_0, \tag{4.1}$$

where d_0 and θ_0 are interplanar distance and X-ray interference angle in stress-free state, respectively.[9]

The relaxation of microstresses results in the emergence of dislocations. Krivoglaz[10] demonstrated that the density of randomly distributed dislocations r is connected with the level of microdistortions via the proportionality coefficient K:

$$\rho = K\varepsilon^2. \tag{4.2}$$

Since the interplanar distance changes continuously in interval $d \pm \Delta d$, the crystal will reflect in the angle range $\theta \pm \Delta\theta$. Thus, we obtain the broadening of the diffraction maximum line by the value:

$$\beta_{\text{MKP}} = 2\Delta(2\theta) = 4\varepsilon \cdot tg(\theta) \tag{4.3}$$

The peak, smearing out, transforms from the δ-like to dome-like (Fig. 4.2c).[10] The defects causing the broadening of diffraction reflexes are referred to class II. Null-dimensional defects are referred to class I: vacancies, impurity, and interstitial atoms.

The broadening of diffraction reflexes can be caused by small dimensions of crystallites being the violation of “ideality” of X-ray interference conditions. The appearance of this defect becomes visible at the sizes of coherent scattering regions (CSR) under 150 nm. Selyakov and Scherrer, independently, obtained the expression for the average maximum smearing from CSR smallness[9]:

$$\beta_{\text{CSR}} = m\frac{\lambda}{D_{hkl}\cos(\theta)}, \tag{4.4}$$

where $m \approx 1$ (the exact meaning depends on the crystal shape and varies in the range between 0.98 and 1.39) and D_{hkl} is the average region size along the normal line to the plane (hkl). The less the D_{hkl} is, the more is the broadening.

The smearing of diffraction reflexes due to microstresses and small sizes of crystallites is called "physical broadening." However, the smearing of ideal δ-like reflex also occurs due to the instrument broadening appearing due to the nonideality of the diffractometer itself. Errors occurring due to the finite width of the counter slit and projection of the tube focus make the greatest contribution to the instrument broadening. Different methods of adjusting the theoretically calculated profile of diffraction pattern to the profile experimentally obtained are used to separate the contributions of physical broadening and instrument broadening. The method of approximation by Gaussian or Cauchy functions and Rietveld method are most widely spread. These methods are realized in the majority of modern hardware and software systems of X-ray structural analysis.

4.2 EXAFS SPECTROSCOPY

Copper is introduced into zinc sulfide as the main electroluminescence activator. Copper can react with sulfur and form the inclusions of Cu_xS phase. At the same time, a lot of heterojunctions of ZnS/Cu_xS, on which the generation and radiative recombination of charge carriers take place during the electric field excitation, are formed in the phosphor volume. X-ray structural analysis allows detecting and attesting the structure of copper sulfide inclusions. Copper can form point defects of penetration, being the luminescence centers. The emission color depends on penetration type of copper and coactivators. Associates (copper in zinc node)–(copper in interstice) produce the emission of blue color, associates (copper in zinc node)–(chlorine in sulfur node)—the emission of green color.[11] Thus, depending on the copper concentration, the predominance of one or another electroluminescence mechanism is possible. Furthermore, it becomes problematic to determine the position of copper atom in the zinc sulfide lattice and its chemical state by the method of X-ray structural state. In this case, it is possible to apply X-ray absorption fine structure (XAFS) methods for investigating the local atomic structure.

Extended X-ray absorption fine structure (EXAFS) spectroscopy based on the investigation is one of the direct investigation methods of local atomic environment. X-ray absorption spectrum represents the measuring of the sample absorption coefficient $\mu(h\nu)$ as the function of energy of incident photons $h\nu$ and possesses sharp absorption edges at energies, at which the excitation of certain core levels becomes accepted (Fig. 4.3).[12] The energies of absorption edges are characteristic for each element, for example, absorption edges K, L_I, L_{II}, and L_{III} correspond to the excitation of electrons from 1s, 2s, 2p1/2, and 2p3/2 states, etc.

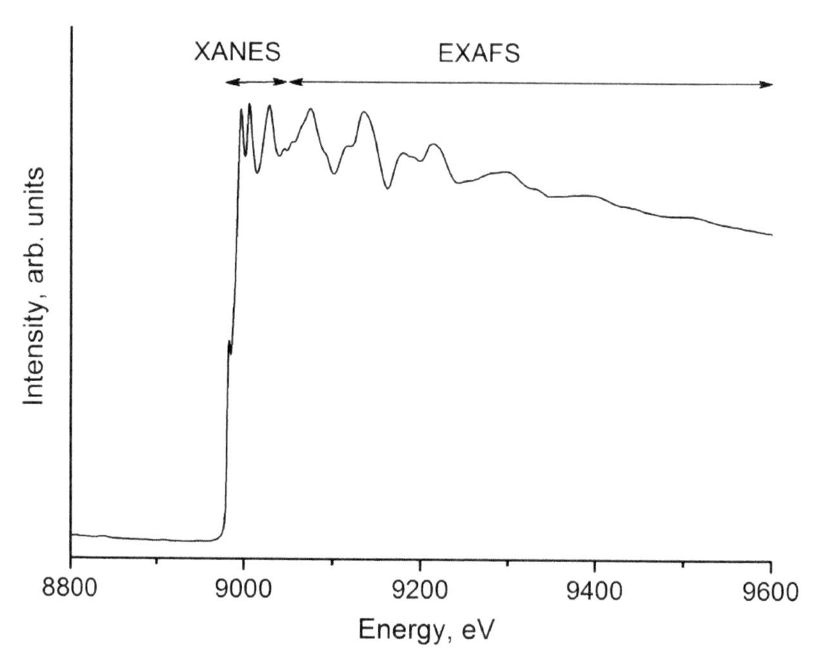

FIGURE 4.3 View of EXAFS spectrum on Cu-K absorption edge and conventional division into two regions: EXAFS and XANES.

Physical basics of EXAFS method can be described in the frameworks of scattering theory. Let us consider an atom bound in a molecule or solid. According to single-electron approximation, the process of X-ray quantum absorption is described with the help of initial state, comprising the incident photon with energy $h\nu$ plus the electron in the main state, and the finite state, comprising the hole plus the electron in the excited state (photoelectron). The initial state is characterized by wave function ψ_i with binding energy E_B, and the finite one—by wave function ψ_f with binding energy E_f.[13] The energy conservation law for this process can be written down as follows:

$$E_{\rm f} = h\nu - E_{\rm B}. \tag{4.5}$$

If the photoelectron is released from 1s level, it behaves as a spherical wave propagating from the excited atom with the wave number as follows:

$$k = \frac{2\pi}{h}\sqrt{2m_0\left(h\nu - E_0\right)}, \tag{4.6}$$

where h—Planck constant and m_0—mass of free electron.[13]

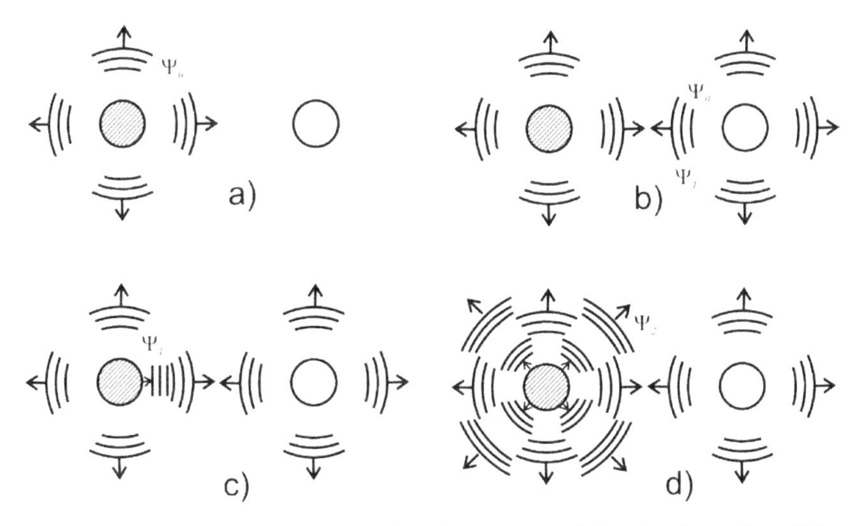

FIGURE 4.4 Formation of the diverging electron wave during the absorption of X-ray radiation in multi-atom system in the single scattering approximation. The hatched circle—absorbing atom, light circle—neighboring atom dissipating the electron wave: (a) Ψ_0 initial electron wave propagating from the absorbing atom; (b) Ψ_1—wave propagating from the neighboring atom when the initial wave Ψ_0 is scattering on it; (c) scattering of wave Ψ_1 on the absorbing center; and (d) its interference with the initial wave Ψ_0 and formation of the resultant wave Ψ_2.

If the atom is bound in the molecule or solid, the photoelectron wave function will be modified by the presence of the potential of surrounding atoms. Such modification can be described as photoelectron scattering (single or multiple) on the atoms of the immediate surrounding (Fig. 4.4).[14]

Then, the wave function of the propagating spherical wave plus the scattered waves corresponds to the finite state. The interference of inversely scattered and initial waves results in the oscillations in the sample $\mu(h\nu)$. These oscillations compose the extended structure having entitled the method, which contains the information on the local geometry of the absorbing atom: lengths of chemical bonds (with the accuracy of up to 0.01 Å for the two nearest coordination spheres[15]) and some angles between them, type of the surrounding atoms, and coordination numbers.[16,17]

EXAFS signal, as a rule, designated as $\chi(k)$, is extracted from the experimentally measured function $\mu(k)$ near the absorption edge (20–300 eV to the edge and 1000–1500 eV after the edge) by subtraction the background (both the pre-edge and atom-like absorption) and normalizing by the value of absorption edge jump to take the sample thickness into consideration. Thus, as a result of the preliminary processing (Fig. 4.5a–c), the normalized oscillation part (NOP) is extracted from the experimentally obtained EXAFS spectra:

$$\chi(k) = \frac{\mu(k) - \mu_0(k)}{\mu_0(k)}, \tag{4.7}$$

where $\mu(k)$ is the above-indicated experimentally measured function, and $\mu_0(k)$ is the atomic absorption coefficient.

Then, in the frameworks of single scattering approximation, NOP can be written down as the photoelectron wave number function:

$$\chi_i(k) = \sum_j \frac{N_j S_0^2(k)}{kR_j^2} \left| f_j^{\mathrm{eff}}(k) \right| e^{-2k^2\sigma_j^2} e^{-2R/\lambda(k)} \sin\left(2kR_j + \phi_{ij}(k)\right), \tag{4.8}$$

where N_j is the number of atoms in j coordination sphere, R_j is the distance from the absorbing atom in i position to the coordination sphere j, $f_j^{\mathrm{eff}}(k)$ is the function of the effective amplitude of the wave scattering, $S_0^2(k)$ is the amplitude diminishing factor due to multiple processes, $\varphi_{ij}(k)$ is the phase shift, σ_j^2 is the Debye–Waller factor describing the change in the radii of coordination spheres due to heat movements and structural disorder, and λ is the length of the photoelectron free path.[12]

All the methods of structural information extraction from EXAFS spectra can be split into two groups[16]: "model-dependent" and "model independent" from the atomic structure of the objects investigated. The method of parametric curve fitting,[18] cumulant method[19] can be ranged in the first group, and Fourier transform,[20] regularization method[21] and method based on the principle of weighted entropy maximum[22]—in the second.

Performing Fourier transforms of $\chi(k)$ function, it is possible to mark the contributions in R-space of different coordination spheres, as demonstrated in Figure 4.5d. Peaks a, b, and c are the contributions of the first, second is the and third coordination spheres, respectively. Fourier transform differs from the function of radial distribution by the fact that it contains the information about both the paired and multi-atomic distribution functions;[23,24] positions of the peaks do not correspond with the actual interatomic distances due to the phase shift $\varphi_{ij}(k)$; the peak shapes are very distorted due to the nontrivial dependence of the back-scattering amplitude on k. Fourier transform can give only the approximate idea about the geometrical structure and its complete interpretation is possible only based on modeling.

One of the possible approaches that allow analyzing the experimental EXAFS spectra is "path fitting," realized, in particular, in software package Ifeffit[25] and corresponding user interfaces Athena and Artemis.[26] All configurations of photoelectron scattering, starting from the absorbing atom, moving to one or several neighboring atoms and returning to the absorbing atom, contribute to EXAFS signal. These configurations are called "scattering paths." The examples of such paths are given in Figure 4.6.

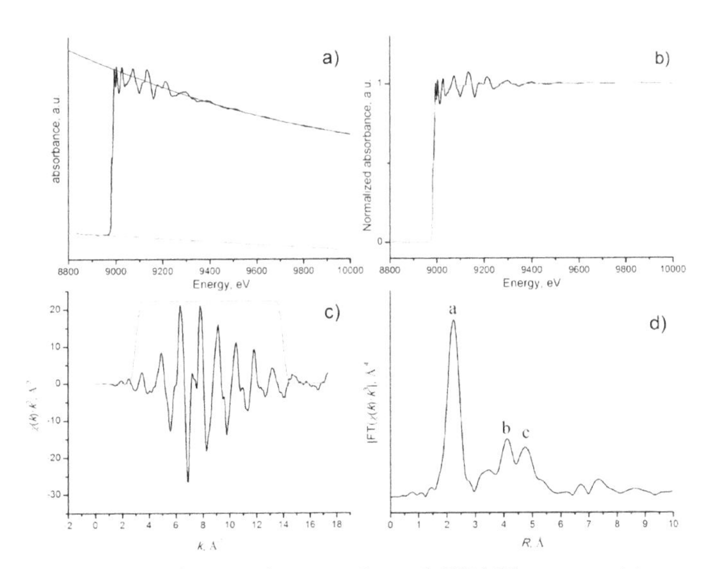

FIGURE 4.5 Stages of processing experimental EXAFS spectra: (a) experimental spectrum, pre-edge background, after-edge background, determination of the absorption jump, and absorption edge; (b) normalized curve of X-ray absorption spectrum, as well as the nonstructured component of spectrum $\mu_0(h\nu)$; (c) oscillating part of spectrum $\chi(k)$ k^3 and window function $W(k)$; and (d) absolute value of Fourier transform of the spectrum oscillating part.

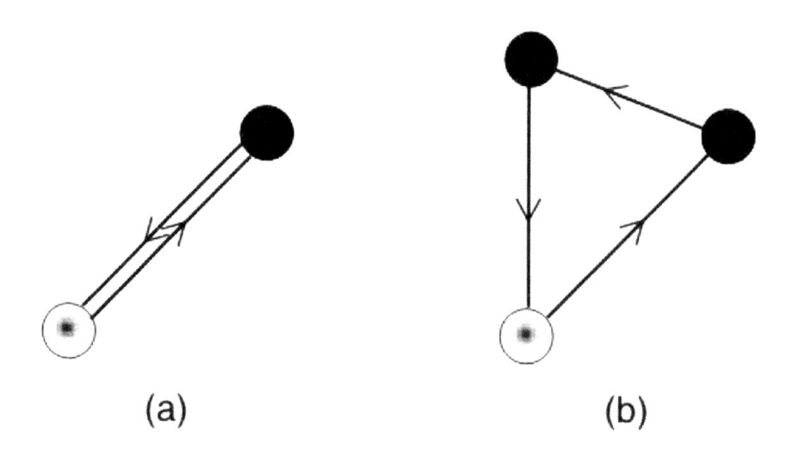

FIGURE 4.6 Schematic view of scattering paths: with the participation of two (a) and three atoms (b). The absorbing atom is indicated as ⊙.

This "model-dependent" approach is based on cumulant splitting of different scattering ways and requires some preliminary knowledge on the system under investigation. The analysis starts with the selection of preliminary structural geometrical model, where the absorbing atom is indicated, as well as its position and type of the neighboring atoms, which need to be considered in the fitting. Crystallographic data, results of molecular mechanics calculations or quantum-chemical optimization serve as the information sources.

The effective scattering amplitudes and phase shifts are calculated in the frameworks of multiple scattering (MS) approach with the help of software Feff.[27] Then, the values of such parameters as interatomic distances R_j, coordination numbers N_j, Debye–Waller factors σ_i^2, and chemical shift ΔE_0 are corrected by minimizing the residual functional "theory-experiment."

In the process of fitting, the total squared deviation for all points of Fourier transform is minimized in the considered region of the spectrum:

$$S = \sum_i \left(F(R)_i^e - F(R)_i^m \right)^2, \tag{4.9}$$

where $F(R)_i^m$ is the Fourier transform of the model spectrum, $F(R)_i^e$ is the Fourier transform of experimental spectrum.

$$F(R) = \frac{1}{2\pi} \int_{k_{min}}^{k_{max}} k^n W(k) \chi(k) e^{2ikR} \, dk \tag{4.10}$$

where k_{min} and k_{max} are the minimum and maximum values of the module of photoelectron wave vector of the oscillating part of experimental spectrum).

The fitting quality is determined by R factor:

$$R = \sum_i \frac{\left[Im\left(F_{exp}(R_i) - F_{theor}(R_i)\right)\right]^2 + \left[Re\left(F_{exp}(R_i) - F_{theor}(R_i)\right)\right]^2}{\left[Im\left(F_{exp}(R_i)\right)\right]^2 + \left[Re\left(F(R_i)\right)\right]^2}, \tag{4.11}$$

The fitting is considered satisfactory if R-factor is less than 0.05.

Such investigation method provides the following opportunities in determining parameters of local atomic structure near the absorbing atom[12]:

(1) determination of radii of coordination spheres with $\Delta R/R \approx 1\%$,
(2) determination of coordination numbers $\Delta N/N \approx 10\%$, and
(3) determination of Debye–Waller factors $\Delta\sigma/\sigma \approx 20\%$.

The main advantage of such approach is the possibility to analyze structural parameters outside the first and second coordination spheres, where the important information is often contained, which is impossible to obtain from the first coordination sphere. Besides, such fitting does not require measuring the reference standard with similar structure and chemical composition.

Software package Iffefit, together with the convenient graphic interface, provides a user with the wide selection of spectrum processing modes. The main shortcomings of the software and the approach itself are the difficulty in analyzing MS ways (actually it is impossible to connect the optimized parameters of MS with the geometric structure parameters),[12] necessity of initial structural model, and strong dependence of the quality of results on the quality of theoretical calculations.[28]

4.3 XANES SPECTROSCOPY

X-ray absorption near-edge structure (XANES), less developed, or practiced than EXAFS, may provide valuable information about the oxidation state, coordination environment, and bonding characteristics of specific elements in a sample. The less common term near-edge X-ray absorption fine structure is used generally in the context of solid-state studies and is synonymous with XANES. The technique in practice requires a mix of qualitative and quantitative analysis to interpret the data and draw conclusions.

XANES is the study of the features immediately before and after the edge, within approximately 1% on either side of the main edge energy. Features include the edge position (a primary indicator of oxidation state), presence/shape of small features just before the main edge ("pre-edge" features), and intensity, number, position, and shape of peaks at the top of the main edge (see Fig. 4.7). The pre-edge feature is weak as it involves a forbidden transition (which is partially allowed due to mixing of p-type impurity), while the first line on the main edge is due to the allowed 1s to 4p bound-state transition.

The power of XANES lies in the sensitivity of edge features to the chemical environment. The sensitivity varies among elements from just detectable to pronounced.

In the XANES regime, the photoelectrons populate unoccupied states above the chemical potential μ. Since—because of the scattered wave—the density of the final states changes, and along with it the charge density, the shape of the absorption edge of the same element is different in different chemical environments. Also, the electron binding (i.e., ionization) energies change for the same reason, which results in shifting of the absorption edge toward lower or higher energies.[29]

As an example, in Figure 4.8, the *K*-edge absorption spectra of zinc oxide, zinc metal, diethyl zinc, and zinc monatomic vapor are shown. Zn vapor is an exceptional case among the shown spectra. While its shape cannot be explained by means of the MS theory, it can be decomposed to contributions of separate excitation channels,[30,31] the shape of which is well known. It has therefore been added for completeness.

At first, we note that the apparent positions of the edges are shifted one to the other. Moreover, the pre-edge resonance that is attributed to the 1s → 4p transition in the case of the atomic spectrum[32] becomes broader and lower quasiatomic resonance (an altered bound state) in Zn–et_2, whereas in Zn metal and Zn oxide, this spectral feature cannot be discerned. Also further on, the differences between individual species are apparent. An electron of low kinetic energy is scattered strongly and possibly many times by the neighboring atoms.

In contrast with single scattering which prevails in the high-energy region, MS depends strongly on angles between scatterers and is thus very sensitive to local environment of the absorbing atom[33]: Figure 4.8 demonstrates strong dependence of the shapes of the spectra on the atoms among which Zn resides.

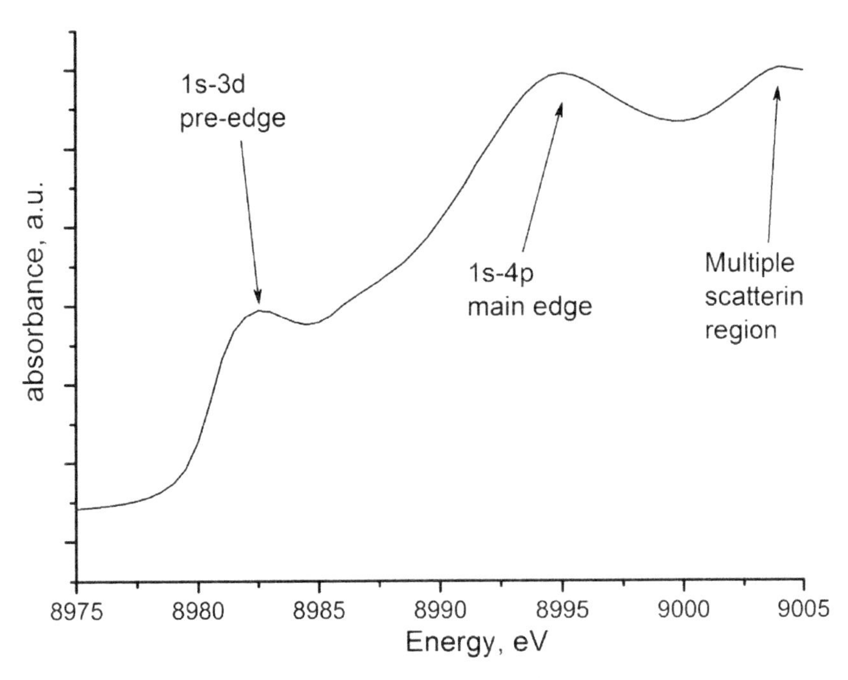

FIGURE 4.7 XANES-spectrum on Cu-*K* absorption edge.

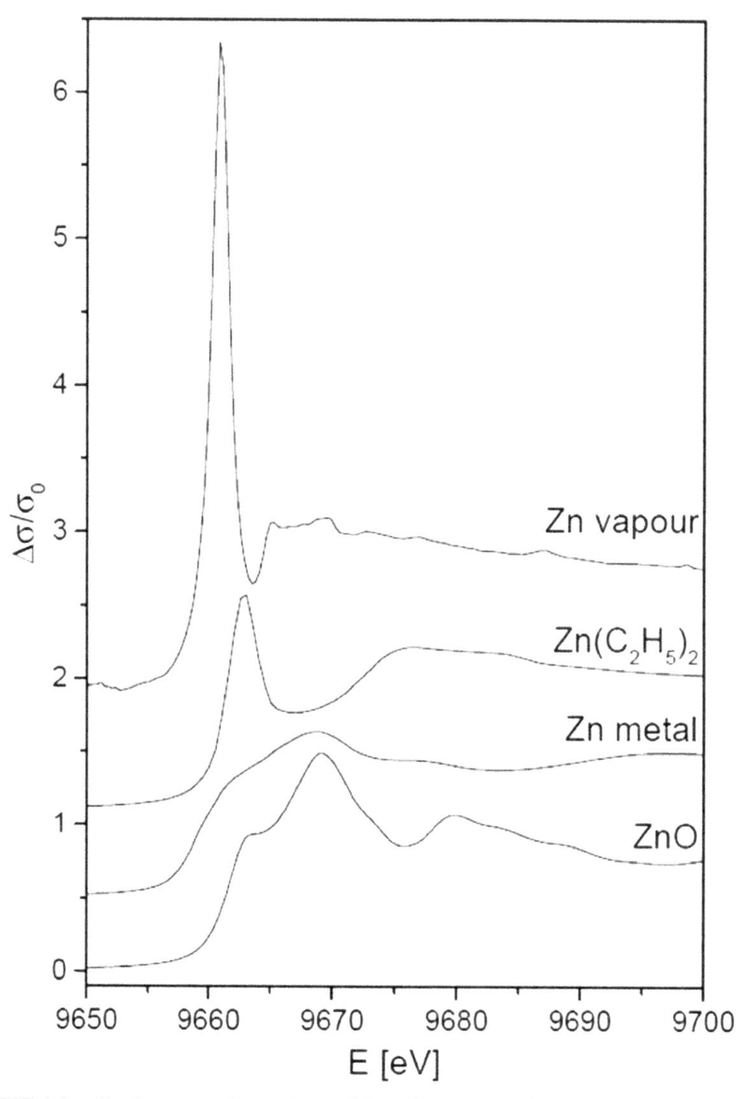

FIGURE 4.8 *K*-edge near-edge regions of Zn oxide, Zn metal, Zn-diethyl, and Zn vapors.

4.4 X-RAY PHOTOELECTRON SPECTROSCOPY

The control of chemical purity is especially important for semiconductors, as even small amount of admixture results in significant changes in electron properties.[34] Secondary ion mass spectrometry is the most sensitive method to determine the composition and amount of admixtures. However, it does not allow to perform a chemical analysis. Therefore, X-ray photoelectron spectroscopy (XPS) is the most popular method among the surface methods. The method is also rather sensitive and allows detecting elements with the concentration up to 0.3% (depending on the atom kind). Besides, the position and shape of the peaks appear to be sensitive to the chemical state. This allows determining chemical bonds, apart from the element composition. The analysis depth is about three monolayers, which makes XPS a surface-sensitive method that is especially important for the investigation of thin films and nanostructures.[35] Apart from the qualitative analysis, semiquantitative and quantitative estimations are possible. A scheme of XPS experiment is presented at Figure 4.9.

XPS can be performed using a commercially built XPS system with an electron energy analyzer. XPS instruments used either a focused 20–500-μm diameter beam of monochromatic Al K_α X-rays, or a broad 10–30-mm diameter beam of nonmonochromatic (polychromatic) Al K_α X-rays or Mg K_α X-rays.

Because the energy of X-ray with particular wavelength is known (for Al K_α X-rays, $E_{ph} = 1486.7$ eV), and because the emitted electrons' kinetic energies are measured, the electron binding energy of each of the emitted electrons can be determined by using an equation that is based on the work of Ernest Rutherford (1914):

where E_b is the binding energy of the electron, E_{ph} is the energy of the X-ray photons being used, E_{kin} is the kinetic energy of the electron as measured by the instrument, and φ is the work function dependent on both the spectrometer and the material. This equation is essentially a conservation of energy equation. The work-function term φ is an adjustable instrumental correction factor that accounts for the few eV of kinetic energy given up by the photoelectron as it becomes absorbed by the instrument's detector. It is a constant that rarely needs to be adjusted in practice.

A typical XPS spectrum is a plot of the number of electrons detected (sometimes per unit time) versus the binding energy of the electrons detected (Fig. 4.10). Each element produces a characteristic set of XPS peaks at characteristic binding energy values that directly identify each element that exists in or on the surface of the material being analyzed. These characteristic spectral peaks correspond to the electron configuration of the electrons within the atoms, for example, 1s, 2s, 2p, 3s, etc. The number of detected electrons in each of the characteristic peaks is directly related to the amount of element within the XPS sampling volume. To generate atomic percentage values, each raw XPS signal must be corrected by dividing its signal intensity (number of electrons detected) by a "relative sensitivity factor" and normalized over all of the elements detected. Since hydrogen is not detected, these atomic percentages exclude hydrogen.

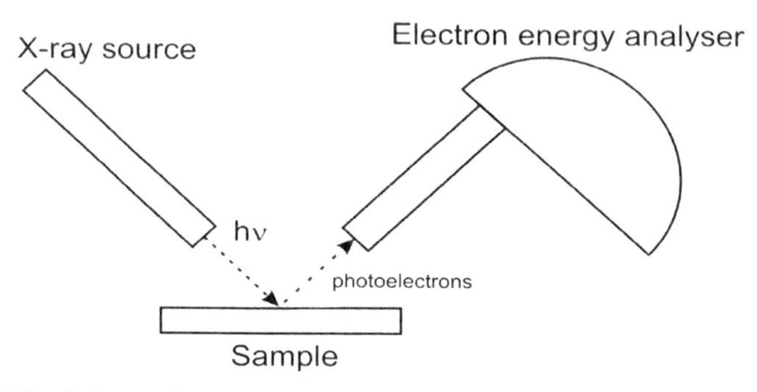

FIGURE 4.9 Scheme of XPS technique.

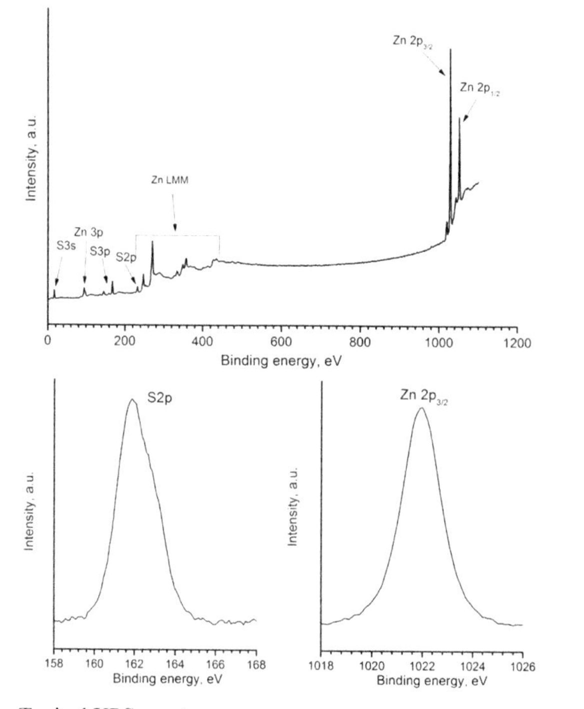

FIGURE 4.10 Typical XPS spectra.

To count the number of electrons during the acquisition of a spectrum with a minimum of error, XPS detectors must be operated under ultrahigh vacuum conditions because electron-counting detectors in XPS instruments are typically 1 m away from the material irradiated with X-rays. This long path length for detection requires such low pressures.

If the measurements are taken on one and the same instrument, in one and the same geometry and with similar parameters of the analyzer, the concentration of element x is expressed as follows:

$$N_x = \frac{I_x}{S_x \sum_i I_i / S_i}, \tag{4.12}$$

where I_x and S_x are the integral intensity and sensitivity coefficient of element x, and the summation is done by all elements in the sample composition.[36]

In general, the values of sensitivity coefficients are determined by the parameters of ionization cross-sections of the corresponding energy levels and can be found in various reference books. Depending on the analyzer type and method of spectrum registration, some deviations can be observed for different instruments.[35]

The correctness of the background component deduction plays the main role in spectrum processing. Linear approximation is the simplest deduction method. In this method, the background emergence nature conditioned by inelastic electron scattering is absolutely not taken into account. This results in errors, especially in the region of large bond energies.

Tougaard algorithm is based on the mechanism of energy losses of electrons on leaving a solid and, therefore, has a certain physical sense. Unfortunately, as a rule, the loss function is frequently unknown. In this case, it is possible to use the universal three-parameter Tougaard function determined by fitting with the help of the least-square method of loss function of noble metals.[37] The universal function has one monotonously descending broad maximum. Such shape does not allow describing some actual structures, for instance, plasmons, characteristic for all metals.[38]

The background in each point in Shirley method is bound with inelestically scattered electrons, which, as assumed, emerge due to scattering the electrons with higher energies. Thus, the background becomes proportional to the integral intensity of electrons with higher kinetic energy. The sensitivity of such method to the occurrence of errors increases with the increase in energy interval, on which the background deduction takes place. The realization of Shirley method assumes some number of iterations; moreover, the increase in this number results in the noise gain in the solution and emergence of artifacts.[38] Nevertheless, this method is one of the most frequently used.

The approximation of photoelectron lines by different functions is used for the interpretation of spectral data. Taking into account that the line shape can often be quite complicated, the mixed Gauss–Lorentz function is most frequently applied for more accurate approximation[39]:

$$f(x)=\frac{\text{Peak height}}{\left[1+\left(M_{\text{mix}}\left(x-x_0\right)^2/\beta^2\right)\right]e^{(1-M_{\text{mix}})\left[\ln 2(x-x_0)\right]^2/\beta^2}}, \quad (4.13)$$

where x_0 is the position of the peak center, β is the parameter, which approximately equals 0.5 of the total width on half-height, M_{mix} is the parameter of function mixing equaled to 1 for purely Lorentz peak and to 0 for purely Gaussian peak.

Thus, it is possible to describe all framing and Auger lines, find the width of the peaks, intensity, and integral intensity. In case of complex chemical composition with overlapping of two and more lines, the knowledge of the shape of reference standard lines allows reducing the spectrum into the components.

KEYWORDS

- **X-ray diffraction**
- **extended X-ray absorption fine structure spectroscopy**
- **X-ray absorption near-edge structure spectroscopy**
- **X-ray photoelectron spectroscopy**
- **chemical composition**

REFERENCES

1. http://www.bourevestnik.ru/products/rentgenodifraktsionnyy-analiz/. (access date 16 Oct, 2017)
2. Kazankin, O. N.; Pekerman, F. M.; Petoshina, L. N. Chemistry and Technology of Luminophors. *Collect. Pap. State Inst. Appl. Chem.* **1960,** *43*, 46–50.
3. Ballentyne, D. H. Electroluminescence—A Disorder Phenomenon. *J. Electrochem. Soc.* **1960,** *107* (10), 807–810.
4. Short, M. A.; Steward, E. G.; Tomlinson, T. B. Electroluminescence in Disordered Zinc Sulphide. *Nature* **1956,** *177*, 240–241.
5. Kazankin, O. N.; Markovsky, L. Ya.; Mironov, I. A.; Pskerman, F. M.; Petoshina, L. N. *Inorganic Luminophors*; Chemistry, Leningrad, 1975; p 118 (in Russian).
6. Krylov, P. N.; Gilmutdinov, F. Z.; Romanov, E. A.; Fedotova, I. V. The Influence of Thermal Annealing on the Optical Properties of Nanocrystalline Zinc Sulfide Films. *Semiconductors* **2011,** *45* (11), 1512–1516.
7. Miloslavsky, V. K.; Naboikina, E. N.; Lebedev, V. P.; Khramtsova, V. I. Optical and Structural Properties of Epitaxial Layers of Zinc Sulfide and Selenide. *Ukrain. Phys. J.* **1969,** *14* (5), 818–825.
8. Krylov, P. N.; Romanov, E. A.; Fedotova, I. V. The Effect of Thermal Annealing on the Structure of Nanocrystalline Zinc Sulfide Films. *Semiconductors* **2011,** *45* (1), 125–129.
9. Usmansky, Ya. S.; Skakov, Yu. A.; Ivanov, A. N.; Rastorguev, L. N. *Crystallography, X-Ray Radiography and Electron Microscopy*; Metallurgy, Moscow, 1982; p 632.
10. Krivoglaz, M. A. *Theory of Scattering X-Rays and Thermal Neutrons by Real Crystals*; Nauka, Moscow, 1967; p 336.
11. Vlasenko, N. A.; Gergel, A. N. On the Mechanism of DC Electroluminescence in *p*Cu,S–*n*ZnS:Mn,Cu,Cl Film Structures. *Phys. Stat. Sol.* **1968,** *26*, K66–K81.
12. Trigub, A. L. Investigation of the Local Atom and Electron Structure of the Complexes of Transition Metals with Porphyrins and Their Analogs by the Methods of X-Ray Absorption Spectroscopy. *Thesis of Candidate of Science in Physics and Mathematics*; Research Center "Kurchatov's Institute", Moscow, 2014; p 141.
13. Lamberti, C. The Use of Synchrotron Radiation Techniques in the Characterization of Strained Semiconductor Heterostructures and Thin Films. *Surf. Sci. Rep.* **2004,** *53*, 1–197.
14. Borovskiĭ, I. B.; Vedrinskiĭ, R. V.; Kraĭzman, V. L.; Sachenko, V. P. EXAFS Spectroscopy: A New Method for Structural Investigations. *Soviet Phys. Uspekhi* **1986,** *149*, 275–324.
15. Kochubey, D. I.; Babanov, Yu. A.; Zamaraev, K. I.; Vedrinski, R. V.; Kraizman, V. L.; Kulipanov, G. P.; Mazalov, L. N.; Skrinski, A. N.; Federov, V. K.; Helmer, B. Yu.; Shuvaev, A. T. *X-ray Spectral Method of Studying the Structure of Amorphous State: EXAFS Spectroscopy*; Nauka, Novosibirsk; 1988; p 306.
16. Valeev, R. G. Nanocomposite Films of Germanium and Gallium Arsenide: Obtaining Methods, Local Atomic Structure, Electrophysical and Photoelectric Properties. *Thesis of Candidate of Science in Physics and Mathematics*; Physical-Technical Institute of UB RAS, Izhevsk, 2003; p 138.

17. Boiko, M. E.; Sharkov, M. D.; Boiko, A. M.; Konnikov, S. G.; Bobyl', A. V.; Budkina, N. S. Investigation of the Atomic, Crystal, and Domain Structures of Materials Based on X-Ray Diffraction and Absorption Data: A Review. *Tech. Phys.* **2015,** *60* (11), 1575–1600.
18. Stern, E. A. Theory of Extended X-Ray Absorption Fine Structure. *Phys. Rev. B* **1974,** *10* (8), 3027–3037.
19. Bunker, G. Application of the Ratio Method of EXAFS Analysis to Disordered Systems. *Nucl. Instr. Method* **1983,** *207* (3), 437–444.
20. Sayers, D.; Stern, E.; Lytle, F. New Technique for Investigation Noncrystalline Structures: Fourier Analysis of the Extended X-Ray Absorption Fine Structure. *Phys. Rev. Lett.* **1971,** *27*, 1204–1207.
21. Tikhonov, A. N.; Arsenin, V. Ya. *Methods of Solving Incorrectly Set Up Problems*; Nauka, Moscow, 1979; p 285 (in Russian).
22. Kuzmin, A. EDA: EXAFS Data Analysis Software Package. *Phys. B* **1995,** *208–209*, 175–176.
23. Aksenov, V. L.; Kuzmin, A. Y.; Purans, J.; Tyutyunnikov, S. I. EXAFS Spectroscopy at Synchrotron-Radiation Beams. *Phys. Particles Nucl.* **2001,** *32* (6), 1–33.
24. Aksenov, V. L.; Koval'chuk. M. V.; Kuz'min, A. Y.; Purans, Y.; Tyutyunnikov, S. I. Development of Methods of EXAFS Spectroscopy on Synchrotron Radiation Beams: Review. *Cryst. Rep.* **2006,** *51* (6), 908–935.
25. Newville, M. EXAFS Analysis Using FEFF and FEFFIT. *J. Synchrotron Rad.* **2001,** *8* (2), 96–100.
26. Ravel, B.; Newville, M. ATHENA, ARTEMIS, HEPHAESTUS: Data Analysis for X-Ray Absorption Spectroscopy Using IFEFFIT. *J. Synchrotron Rad.* **2005,** *12* (4), 537–541.
27. Rehr, J. J.; Kas, J. J.; Vila, F. D.; Prange, M. P.; Jorissen, K. Parameter-Free Calculations of X-Ray Spectra with FEFF9. *Phys. Chem. Chem. Phys.* **2010,** *12*, 5503–5513.
28. Schnohr, C. S.; Ridgway, M. C., Eds. *X-Ray Absorption Spectroscopy of Semiconductors*; Springer, Berlin, 2015; p 361.
29. http://sabotin.ung.si/~arcon/xas/xanes/xanes-theory.pdf. (accessed Oct 11, 2017)
30. Breinig, M.; Chen, M. H.; Ice, G. E.; Parente, F.; Crasemann, B.; Brown, G. S. Atomic Inner-shell Level Energies Determined by Absorption Spectrometry with Synchrotron Radiation. *Phys. Rev. A* **1980,** *22*, 520.
31. Teodorescu, C. M.; Karnatak, R. C.; Esteva, J. M.; El Afif, A.; Connerade, J.-P. Unresolvable Rydberg Lines in X-Ray Absorption Spectra of Free Atoms. *J. Phys. B* **1993,** *26*, 4019–4039.
32. Miheliˇc, A.; Kodre, A.; Arˇcon, I.; Padeˇznik Gomilˇsek, J.; Borowski, M. A Double Cell for X-Ray Absorption Spectrometry of Atomic Zn. *Nucl. Instrum. Meth. B* **2002,** *196*, 194–197.
33. Koningsberger, D. C.; Prins, R., Eds. *X-Ray Absorption—Principles, Applications, Techniques of EXAFS, SEXAFS and XANES*; John Wiley and Sons, Hoboken, NJ, 1988.
34. Shalimova, K. V. *Physics of Semiconductors*. K. V. Shalimova, Moscow: Energy, 1976; p 416.

35. Briggs, D.; Sikh, M. P. *Surface Analysis by Auger and X-Ray Photoelectron Spectroscopy*. Moscow: Mir, 1987; p 600 (Translated from English).
36. Pratton, M. *Introduction to Surface Physics*. Regular and Chaotic Dynamics, Izhevsk, 2000, p 256.
37. Tougaard, S. Universal Inelastic Electron Scattering Cross-Sections. *S. Tougaard, Surf. Interface Anal.* **1997,** *25*, 137–154.
38. Jo, M. Direct Simultaneous Determination of XPS Background and Inelastic Differential Cross Section Using Tougaard's Algorithm. *Surf. Sci.* **1994,** *320*, 191–200.
39. Zakhvatova, M. V.; Gilmutdinov, F. Z.; Surnin, D. V. Allowance for the Background Component in X-Ray Photoelectron and Auger Electron Spectroscopy. *Phys. Met. Met. Sci.* **2007,** *104* (2), 166–171.

CHAPTER 5

Methods of Optical Spectroscopy

ABSTRACT

This chapter is dedicated to a detailed description of the methods of optical spectroscopy that we use in optical investigations. Monochromator on diffraction grids and the operating principle on these grids have been considered. The technique of determination of the band gap width of semiconductors by the spectra of optical transmission is presented. The main illuminating characteristics of electroluminescent emitters and devices and the methods for measuring electroluminescent emitter radiation characteristics are described.

5.1 MONOCHROMATOR ON DIFFRACTION GRIDS

The absorption spectral analysis is the direct and simplest method of investigating optical properties and band structure of semiconductors and dielectrics. Spectral devices or spectrometers are applied for the investigations, that is, optical instruments used to analyze the spectra of electromagnetic field of the optical range. In this case, under spectrum, we understand the dependence of light intensity on the wavelength $I(\lambda)$ or energy of photons $I(h\nu)$. Spectrometers can be divided into two main groups: dispersive and non-dispersive. In dispersive spectrometers, the spatial separation of light of different wavelengths with the help of the dispersive elements is carried out. Spectral prisms (the phenomenon of refraction index dispersion is used) or diffraction grids (the phenomenon of light dispersion on periodical structure is used) usually serve as dispersive elements.

Let's consider the monochromator operating on principle of diffraction grids, which we used in optical investigations. The simplest diffraction grid consists of transparent regions (slits) separated by nontransparent gaps. The parallel beam of the investigated light is directed at the grid with help of collimator. The observation is performed in the focal plane of the lens placed behind the grid (Fig. 5.1).

In each point, P, on the screen, the beams—which were parallel before the lens and propagated under the certain angle θ to the direction of the down coming wave—will be collected in the focal plane of the lens. The oscillation in point P results from the interference of secondary waves arriving in this point from different slits. To reach the interference maximum in point P, the path difference Δ between the waves emitted by the neighboring slits should equal the integer number of wavelengths:

$$\Delta = d_g \sin\theta_m = m\lambda \ (m = 0, \pm 1, \pm 2, \ldots) \tag{5.1}$$

Here, d_g = grid period and m = integer number, which is called "the diffraction order."

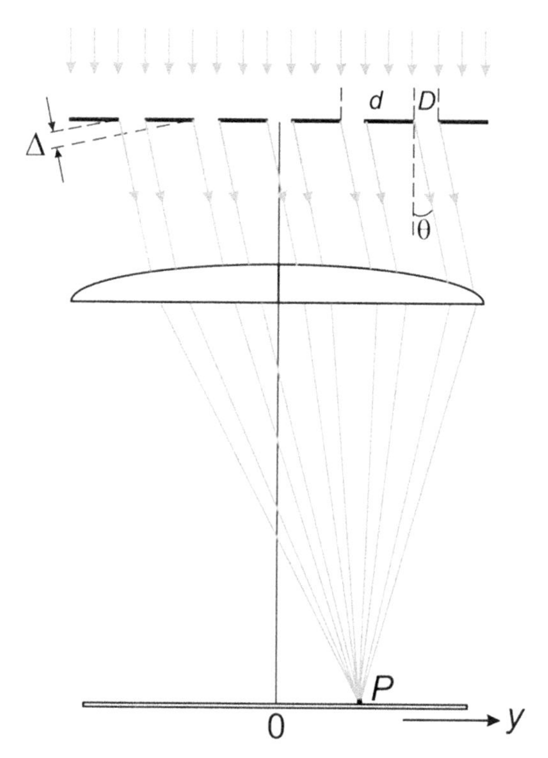

FIGURE 5.1 Optical scheme of monochromator on diffraction grids.

The so-called "main peak wavelengths" of the diffraction pattern are located in those screen points, for which this condition is fulfilled.

In the focal plane of the lens, the distance y_m from the maximum of zero order ($m = 0$) to the maximum of m order at small diffraction angles equals $y_m = m\frac{\lambda}{\alpha}F$, where F = focal distance.

Based on the diffraction grid formula, the position of main maximums (except for the zero one) depends on the wavelength λ. Therefore, the grid can disperse light into the spectrum, that is, it is a spectral instrument. If nonmonochromatic light gets onto the grid, the spectrum of the investigated light emerges in each diffraction order (i.e., with each m value), which is more, the violet part of the spectrum is placed closer to the maximum of zero order. Figure 5.2 demonstrates the spectra of different orders for white light. The maximum of zero order remains achromatic.

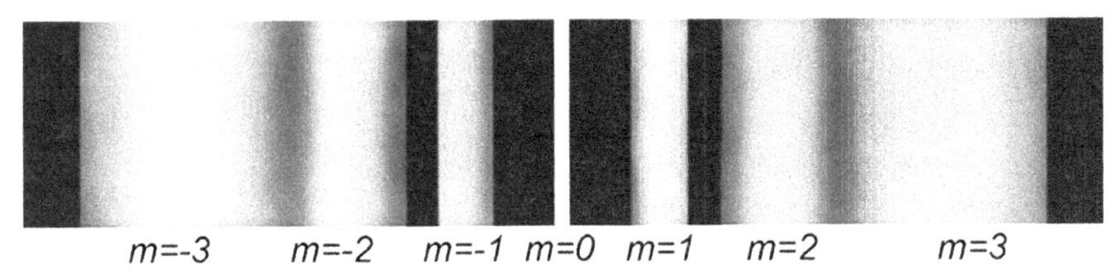

FIGURE 5.2 (See color insert.) White light dispersion into the spectrum with the help of diffraction grid.

With the help of diffraction grid, it is possible to perform very precise measurements of the wavelength. If the grid period d is known, the length is found measuring the angle θ_m of the corresponding direction at the selected line in the spectrum of m order. In practice, the spectra of second or third order are usually used.

The spectrometric complex based on monochromator monochromator on diffraction grid (MDG-41) (Experimental Designing Bureau "Spektr," Saint-Petersburg) was applied to measure the spectra of optical transmission of the samples obtained in this work. The optical scheme of the measurements is demonstrated in Figure 5.3. The sample investigated was placed before the input slit.

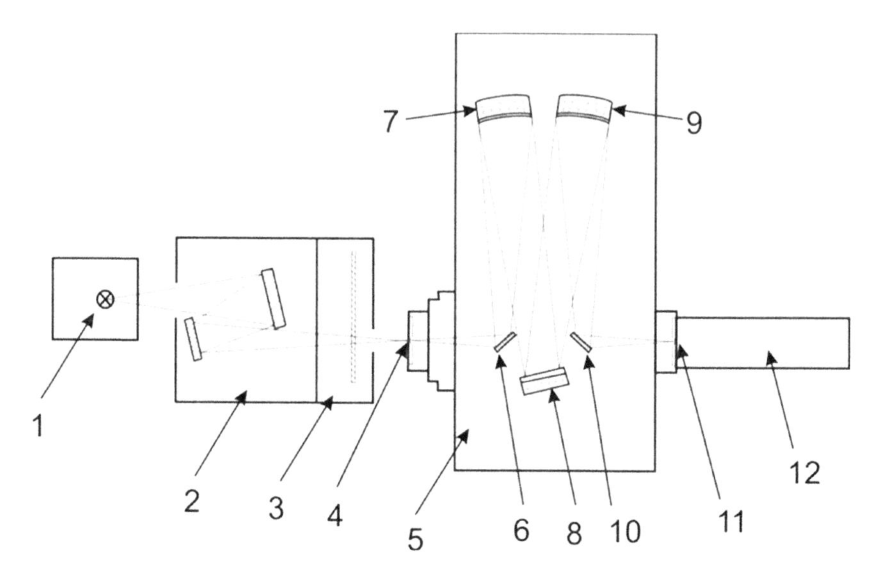

FIGURE 5.3 Optical scheme of the measurements of transmission spectra. 1—light source; 2—mirror condenser; 3—color filter unit; 4—input slit of the monochromator; 5—monochromator body; 6, 10—flat rotating mirrors; 7, 9—parabolic mirrors; 8—diffraction grid; 11—output slit of the monochromator; 12—radiation detector.

5.2 DETERMINATION OF THE BAND GAP OF SEMICONDUCTORS BY THE SPECTRA OF OPTICAL TRANSMISSION

The material band gap can be calculated directly from its optical properties. Transmission and absorption coefficients for absorbing thin films are bound by the following correlation[1–6]:

$$T_{\mathrm{tr}} = (1 - R_{\mathrm{ref}})^2 e^{-\alpha_{\mathrm{ab}} d}, \tag{5.2}$$

where T_{tr} and R_{ref} = transmission and reflection coefficients, respectively,
α_{ab} = absorption coefficient,
d = film thickness.

The expression (5.2) is true, if the retraction index n of the film is more than that of the template, and it allows calculating the absorption spectrum (λ) of the material by experimentally measured transmission $T(\lambda)$ and reflection $R(\lambda)$ spectra:

$$\alpha_{ab} = -\frac{1}{d}\ln\frac{T_{tr}}{(1-R_{ref})^2} \quad (5.3)$$

The optical absorption is conditioned by electron transitions between energy layers. Thus, the band structure can be analyzed by the absorption edges present on the spectrum. The long-wavelength absorption edge of semiconductors and dielectrics is connected with electron transition from the valence region to the conduction region that allows finding the band gap E_g. The correlation between the energy of the band gap E_g and absorption coefficient is set up by the following formula[7]:

$$\alpha_{ab} = \left(\frac{A}{h\nu}\right)(h\nu - E_g)^2, \quad (5.4)$$

where A = Tauc constant,[8]
$h\nu$ = photon energy.

In practice, the graph of dependence of $(\alpha h\nu)^2$ on $h\nu$ (so-called Tauc coordinates) is plotted to determine E_g by the absorption spectrum. The band gap width in the intersection point of the straight line and abscissa axis is found by the edge linear extrapolation (Fig. 5.4).

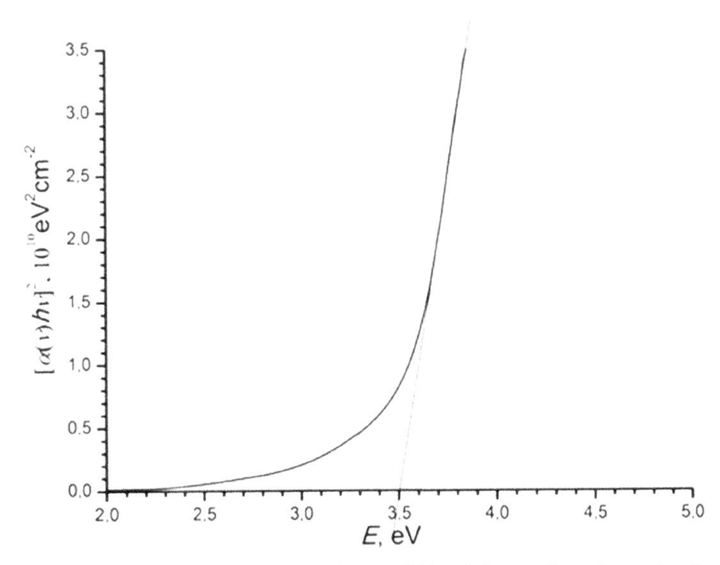

FIGURE 5.4 Determination of the band gap width of the semiconductor by the optical absorption spectra.

5.3 MAIN ILLUMINATING CHARACTERISTICS OF ELECTROLUMINESCENT EMITTERS

As mentioned earlier section(s), electroluminescence emerges under the action of electric field, which changes the potential or kinetic energy of electrons in a solid. During the crystal electroluminescence, the electric energy immediately transforms into the light energy. Moreover, the efficiency coefficient of this process, in some cases, can be as high as 100%.

The luminescence color of electroluminescent light sources (ELS) depends on the phosphor material, nature and concentration of the activator (or several admixtures introduced simultaneously), and conditions of sample excitation. Besides, the emission spectrum can be changed by introducing additional metal or dielectric interlayers to achieve the emission interference.

ELS, in which ZnS is doped by manganese or rare-earth elements (REE) and luminescence, has intracenter character that possess the best characteristics. The introduction of Mn results in the emergence of the band with the maximum of about 580 nm (orange line in Fig. 5.5). Since Mn interaction with ZnS grid is stronger than that of REE, the half-width of Mn band is greater in the spectrum.

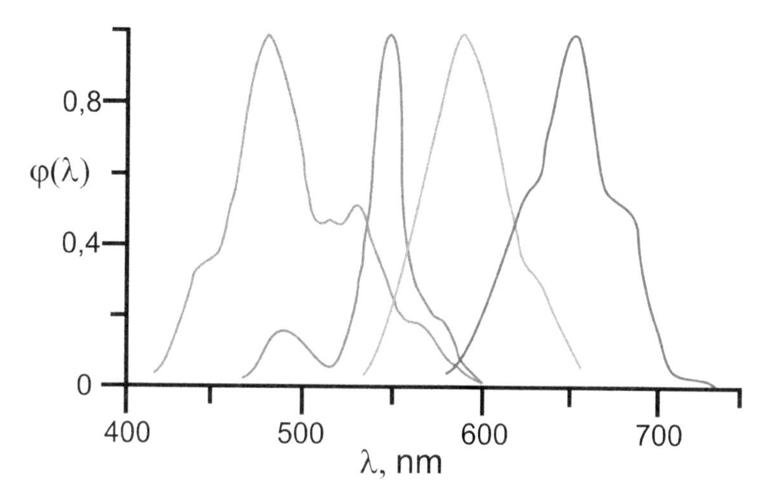

FIGURE 5.5 (See color insert.) Spectral characteristics of electroluminescence (in relative units) of the films based on Ref. [9]: *SrS:Ce (blue line), ZnS:TbF$_3$ (green line), ZnS:Mn (orange line) and CaS:Eu,Ce (red line).*

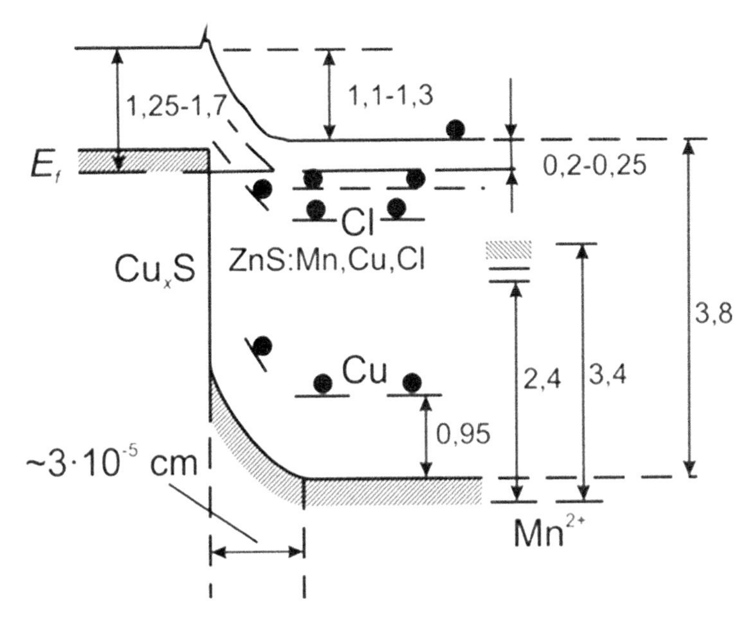

FIGURE 5.6 Scheme of heterojunction between copper and zinc sulfides. E_F = Fermi level. Digits near the arrows indicate energy in *eV*.
Source: Adapted with permission from Ref. [10].

The luminescence of Mn^{+2} ions is caused by the transitions of electrons from the first excited state to the main one (Fig. 5.6). In this state (^6S), all five *d* electrons have parallel spins; in the first excited state ^4G, the spin of one of the electrons is anti-parallel to the others. The symmetry decreases in the luminescence center surroundings (with the change in the synthesis conditions, formation of the charged defects in the structure of Cl^- type in S^{-2} node) resulting in level splitting.

The exchange interaction between the neighboring ions of Mn^{+2} contributes to the emergence of additional maximums ZnS:Mn spectrum. Therefore, apparently, the concentration increase of manganese results in the appearance of additional maximums in ZnS:Mn spectrum near $\lambda = 650$ nm (red color) and mass content of Mn $\geq 10\%$ near $\lambda = 775$ nm (infrared light), whose values, however, are significantly less than the maximum value at $\lambda = 580$ nm (yellow–orange range of the spectrum).

The brightness of ELS luminescence depends on many factors: excitation conditions (amplitude and shape of voltage U applied to the structure, its frequency f, temperature T at which the emitter operates), characteristics of the sample (concentration of admixtures in phosphor layer, material of dielectric layers, thickness of layers), and so on.

The typical volt-brightness characteristic of ELS with two dielectric films has three regions (Fig. 5.7, red curve). In the first, its slope n, characterized by the degree index of function $L \sim U^n$, gradually increases (here $n = 10 \div 15$), it is maximum in the second region ($n \geq 30$), and in the third region brightness L only slightly depends on U.

When explaining this characteristic, it should be kept in mind that it is determined by the dependencies of L on active current I and quantum output of the excitation process of radiation centers N_q: $L \sim IN_qP_{rad}$; it is assumed that the probability of radiation transitions P_{rad} does not depend on U.

The threshold voltage of lighting U_T, under which we understand the voltage corresponding to the definite (threshold) luminescence brightness (most frequently—the brightness of 10 ft-lambert or 34 cd/m^2, ELS production, nature of dielectric layers, activator concentration in phosphor layer).

Despite of the fact that it is possible to obtain structures with U_T~60 V, the threshold voltage of lighting is usually above 100 V.

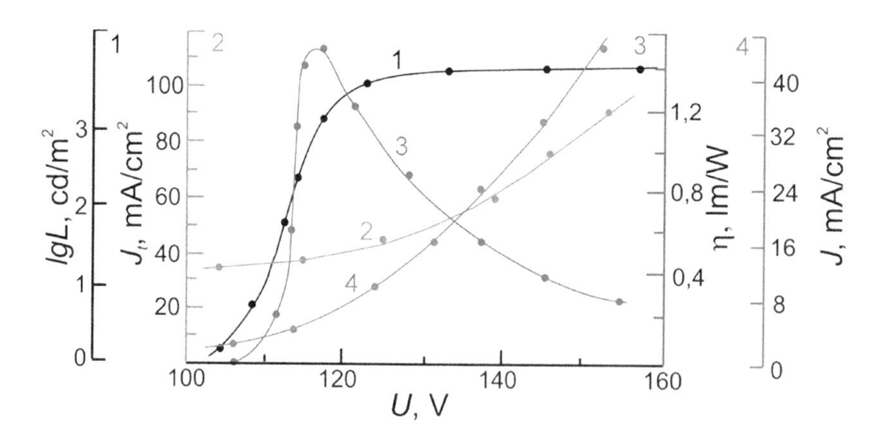

FIGURE 5.7 Dependence of main parameters of ELS with two dielectric layers on the applied voltage U.[11] Black curve—brightness, green curve—density of active current J, blue curve—density of total current J_t, red curve—luminary efficiency η.

5.4 DEVICES AND METHODS FOR MEASURING ELS RADIATION CHARACTERISTICS

The laboratory stand, representing the transparent base with spring-loaded contacts, was produced to investigate electric and light emitting characteristics of ELS. The stand external view is demonstrated in Figure 5.8. Figure 5.9 demonstrates the scheme of ELS prototypes switching to AC voltage source.

FIGURE 5.8 Stand for electric and light emitting investigations of ELS prototypes.

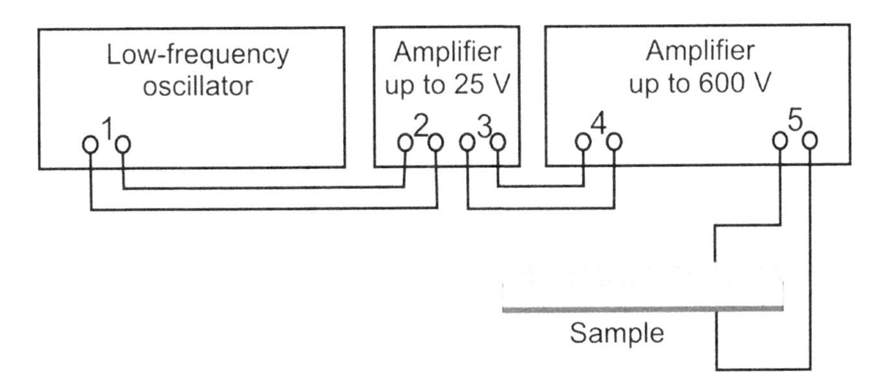

FIGURE 5.9 Scheme of ELS prototypes switching to AC voltage source.

Low-voltage (up to 5 V) signal of rectangular or sinusoidal shape is transmitted from the output 1 (Fig. 5.9) of generator GZ-112 (Fig. 5.10a) to input 2 of pre-amplifier GZ-112/1 (Fig. 5.10b), which boosts the signal up to 25 V. From the pre-amplifier output 3, the signal is transmitted to the input 4 of final power amplifier (Fig. 5.10c), which allows having at the output 5 the potential difference up to 600 V with the frequency set up by the generator. It should be pointed out that the final power amplifier has some voltage dependence on frequency: till the frequencies of about 20 kHz, the maximum voltage increases linearly with the frequency from 250 to 600 V, and then it decreases linearly. Based on this, we selected the frequencies and voltages, at which the radiation spectra, volt-brightness, and frequency-brightness characteristics of the samples were investigated.

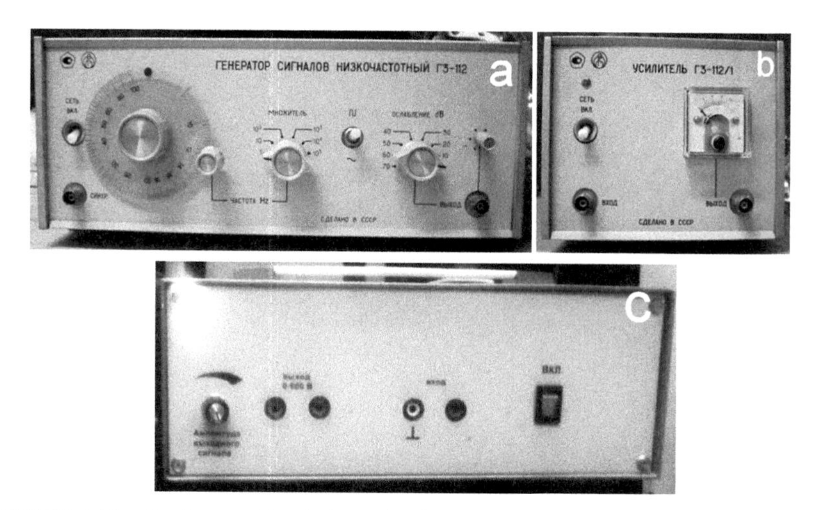

FIGURE 5.10 Devices used for the investigation of ELS characteristics. (a) low-frequency oscillator, (b) amplifier up to 25 V, and (c) amplifier up to 600 V.

Radiation characteristics—intensity and wavelength—were investigated with the help of spectrometer based on monochromator MDR-41 (Fig. 5.3) in the wavelength band from 250 up to 700 nm. The sample was placed immediately in front of the monochromator input slit. The availability of radiation components with different wavelengths was determined while dispersing the spectrum into components of Gaussian curve following the Newton method.

Since the spectrometer registers the number of photons, the brightness was computed from the assumption that the light intensity was measured from the spot light source. The spectrometer really registers the light flux from the area corresponding to geometric parameters of the monochromator input slit. Then, considering that the spot source is placed into transparent homogeneous medium, not only the light flux Φ emitted by the source at some time moment remains constant at any distances from it, but also the light flux $d\Phi$ within any solid angle $d\Omega$ emitted from the source.

As the solid angle $d\Omega$ is not connected with distance r from the source, the source light power F_1 also does not depend on it. The light intensity I_1 at the distance r is found dividing $d\Phi$ by the area $ds=r^2 d\Omega$ of perpendicular section of the elementary beam of the rays considered. This provides

$$I_1 = F_1/r^2 \tag{5.5}$$

The light flux falling on the area unit of the lighted surface is called the illumination (or illumination brightness) of this surface. Let the beam fall from the spot source at the angle θ to the normal line of the illuminated surface. Then $d\Phi = F_1\, d\Omega = (F_1\, ds\cos\theta)/r^2$. Having divided by the surface area ds, we have

$$E_{ill}=(F_1 \cos\theta)/r^2=I_1 \cos\theta. \tag{5.6}$$

In our case, the beam fall along the normal line to the surface, that is, $\cos\theta = 1$. Consequently, the illumination brightness will equal the intensity of the emitted light, that is, the number of photons registered by the spectrometer detector.

KEYWORDS

- **optical absorption**
- **diffraction grid**
- **band gap**
- **electroluminescence**
- **optical spectroscopy**

REFERENCES

1. Swanepoel, R. Determination of the Thickness and Optical Constants of Amorphous Silicon. *J. Phys. E* **1983,** *16* (12), 1214–1221.
2. Sánchez-González, J.; Díaz-Parralejo, A.; Ortiz, A. L; Guiberteau, F. Determination of Optical Properties in Nanostructured Thin Films Using the Swanepoel Method. *Appl. Surf. Sc.* **2006,** *252* (17), 6013–6017.
3. Ilican, S.; Gaglar, M.; Gaglar, Y. Determination of the Thickness and Optical Constants of Transparent Indium-Doped ZnO Thin Films by the Envelope Method. *Mater. Sci. Poland* **2007,** *25*, 709–717.
4. Gumus, C.; Ozkendir, O. M.; Kavak, H.; Ufuktepe, Y. Structural and Optical Properties of Zinc Oxide Thin Films Prepared By Spray Pyrolysis Method. *J. Optoelectron. Anv. Mater.* **2006,** *8*, 299–303.
5. Brus, V. V.; Kovalyuk, Z. D.; Maryanchuk, P. D. Optical Properties of TiO_2–MnO_2 Thin Films Prepared by Electron Beam Evaporation. *Tech. Phys.* **2012,** *57* (8), 1148–1151.
6. Vasilevski, M. A.; Konoplev, G. A.; Panov, M. F. *Optical-physical Methods of Investigations of Materials and Thin-Film Structures;* SPbGETU "LETI": Saint-Petersburg, 2011; p 56 (In Russian).
7. Fahrenbruch, A. L.; Bube, R. H. *Fundamentals of Solar Cells;* Academic Press, Cambridge, USA, 1999; p 850.
8. Tauc, J.; Grigorovichi, R.; Vancuet, A. Optical Properties and Electronic Structure of Amorphous Germanium. *Phys. Stat. Sol.* **1966,** *15*, 627–637.
9. Vlasenko, N. A. Electroluminescent Films as a Light Emitting Sources for Optoelectronics. *Semiconduct. Techniq. Microelect.* **1973,** *3*, 93–101 (In Russian).
10. Vlasenko, N. A.; Gergel, A. N. On the Mechanism of DC Electroluminescence in pCu,SnZnS:Mn,Cu,Cl Film Structures. *Phys. Stat. Sol.* **1968,** *26*, K66—K81.
11. Vlasenko, N. A.; Kurilenko, B. V.; Cirkunov Yu. A. *Electroluminescent Light Sources and Their Application;* Znanie: Kiev, 1981; p 24 (In Russian).

CHAPTER 6

Theoretical Models for Investigating The Processes of Nanofilm Deposition onto Porous Templates of Aluminum Oxide

ABSTRACT

This chapter studies the mathematical models of nanofilm formation on porous templates of aluminum oxide. The equations describing the method of molecular dynamics, as well as the equations of nanoparticle movement in mesic media are demonstrated. The periodic boundary conditions and their influence on the system modeled are formulated. The potentials of different interaction types emerging in the system modeled are described. Different numerical algorithms for solving the modeling problem of nanofilm coating formation are given. The formulas for calculating thermodynamic parameters of the system modeled are analyzed.

6.1 MODELING METHODS OF NANOFILM FORMATION

The following modeling methods were used to solve the problem of the formation of nanofilm coatings, nanostructure formation, their movement and interaction, as well as the processes of nanocomposite formation:

- ab initio method;
- semiempirical methods;
- method of molecular dynamics; and
- method of nanostructure movement in mesic media.

6.1.1 Ab Initio Method and Semiempirical Methods

The apparatus of quantum mechanics is used in ab initio method and it is based on solving the Schrödinger equation. Ab initio method is detailed in the works on quantum mechanics.[1–4] When applying this method, the complete electron and atom structure of the objects (atoms, molecules, and ions) is studied and the detailed configuration of all electron clouds is taken into account. In general, the Schrödinger equation does not have analytical solution, and this problem is solved with the help of numerical methods.

The advantage of ab initio method is the fact that it is not necessary to know any empiric parameters to calculate, for example, force and length of individual bonds, angle values, etc. To conduct a theoretical study, it is enough to know the chemical formula of the studied material as the initial data. If the chemical elements in the material are connected, it is necessary to specify the order of their connection and the number of connections. Owing to the detailed account of electron cloud configuration and atomic structure of the objects investigated, ab initio method is more precise in comparison with other methods. However, due to the fast increase in the complexity of solution search with the increased number of the system elements, the possibilities of quantum-mechanical calculations are limited by the level of computer development.

In quantum mechanics, the system state is described by the wave function of coordinates $\Psi(\bar{\mathbf{r}})$, whose squared absolute value determines the distribution of probabilities coordinate values: $|\Psi(\bar{\mathbf{r}})|^2 d\bar{\mathbf{r}}$ = the possibility that the system will be in the volume element $d\bar{\mathbf{r}}$. The quantum-mechanical methods used in the work are based on the solution of the stationary Schrödinger equation[1]

$$\hat{\mathbf{H}}\Psi = E\Psi, \tag{6.1}$$

where Ψ = all-electron wave function of the system,
E = total energy of the system, and
$\hat{\mathbf{H}}$ = Hamiltonian of the system consisting of the total of operators of kinetic E and potential U energies.

For multi-electron systems, it is impossible to find the precise solution of the Schrödinger equation; therefore, different methods of approximate solution are used. In the system consisting of nuclei N_n and electrons N_e, the Schrödinger equation will contain the variables $3(N_n + N_e)$ in the form of spatial coordinates. In nonrelativistic approximation, Hamiltonian operator of the system will be as follows[2]:

$$\hat{\mathbf{H}} = E_n + E_e + U_n + U_e + U_{ne} \tag{6.2}$$

The operator of kinetic energy of the nuclei is found by the formula

$$E_n = -\sum_{p}^{N_n} \frac{\nabla_p^2}{2m_p}, \tag{6.3}$$

where m_p = mass of the nucleus p,

∇ = Laplace operator, $\nabla^2 = \frac{\partial^2}{\partial x^2} + \frac{\partial^2}{\partial y^2} + \frac{\partial^2}{\partial z^2}$.

The following formula is used to find the operator of kinetic energy of the electrons:

$$E_e = -\frac{1}{2}\sum_{i}^{N_e} \nabla_i^2 \tag{6.4}$$

The potential energy of internuclear repulsion is as follows:

$$U_n = -\sum_{p>l}^{N_n} \frac{q_p q_l}{R_{pl}}, \tag{6.5}$$

where q_p, q_l = charges of the nuclei p and q,

R_{pl} = distance between the nuclei p and q.

The operator of interelectron interaction is calculated as follows:

$$U_e = \sum_{i>j}^{N_e} \frac{1}{r_{ij}}, \tag{6.6}$$

where r_{ij} = distance between the electrons i and j.

The potential energy of attraction of nuclei and electrons is found by the formula

$$U_{ne} = -\sum_{i}^{N_e}\sum_{p}^{N_n}\frac{q_p}{r_{ip}}, \tag{6.7}$$

where q_p = charge of the nucleus p,

r_{ip} = distance between the nucleus p and electron i.

In general, the complete wave function of the system $\Psi(\bar{\mathbf{R}},\bar{\mathbf{r}})$ depends on the coordinates of the nuclei $\bar{\mathbf{R}}$ and coordinates of electrons $\bar{\mathbf{r}}$; consequently, the Schrödinger equation will be written down as follows: $\hat{\mathbf{H}}\Psi(\bar{\mathbf{R}},\bar{\mathbf{r}}) = E\Psi(\bar{\mathbf{R}},\bar{\mathbf{r}})$. Since the nuclei has a large mass in comparison with electrons and they move much slower, Born–Oppenheimer approximation allows dividing the variables of electrons and nuclei and solving the Schrödinger equation separately for nuclear and electron systems. The nuclei move slowly entraining light electrons, and electrons produce the averaged force field, in which the nuclei move.

The electron Hamiltonian operator describing the movement of electrons in the field of fixed nuclei is found by the formula $\hat{\mathbf{H}}_{\text{elec}} = E_e + U_e + U_{ne}$. The corresponding Schrödinger equation for electron structure will be written down as follows[2]

$$\hat{\mathbf{H}}_{\text{elec}}(\bar{\mathbf{r}})\Psi_{\text{elec}}(\bar{\mathbf{r}}) = E_{\text{elec}}(\bar{\mathbf{R}})\Psi_{\text{elec}}(\bar{\mathbf{r}}), \tag{6.8}$$

where $\hat{\mathbf{H}}_{\text{elec}}(\bar{\mathbf{r}})$ = electron Hamiltonian depending only on coordinates of the electrons,

$\Psi_{\text{elec}}(\bar{\mathbf{r}})$ = wave multi-electron function.

Solving this equation with fixed positions of the nuclei, the dependence of the total energy of electron subsystem upon the nuclei position is calculated.

The solution of nuclear subsystem is found based on the solution of the Schrödinger equation. The complete Hamiltonian operator will be written down as follows:

$$\hat{\mathbf{H}}(\bar{\mathbf{R}}) = E_n(\bar{\mathbf{R}}) + \hat{\mathbf{H}}_{\text{elec}}(\bar{\mathbf{r}}) + U_n(\bar{\mathbf{R}}) = E_n(\bar{\mathbf{R}}) + E_{\text{elec}}(\bar{\mathbf{R}}) + U_n(\bar{\mathbf{R}}).$$

In Hartree–Fock method, the wave function of the system of electrons and nuclei Ψ, describing the system state is presented as the determinant $\Psi = \frac{1}{\sqrt{N!}}\det\{\Psi_i\}_{i=1}^{N}$, composed of separate wave functions of atoms within the system, which, in turn, are given as the linear combination of finite amount of basis states[3,4]:

$$\Psi_i = \sum_{v} c_{vi} \varphi_v \tag{6.9}$$

The selection of basic atomic functions is an important task as it determines the accuracy of the wave function approximation expansion. Different sets of functions are selected as the basis: Gaussian orbitals, plane waves, and Slater orbitals.

Thus, the solution of the Schrödinger equation in Hartree–Fock method comes down to searching the coefficients c_{vi}, dependable on spatial coordinates and defining the contribution of the corresponding basis functions to the system energy. c_{vi} can be found based on variation method whose principle consists in the selection of trial wave functions and further correction of coefficients c_{vi} to minimize the system energy.

Although ab initio method requires considerable number of computational resources for its calculations, it is widely spread due to high accuracy of the results. The application of this method is especially convenient to solve problems on geometry optimization, when the equilibrium configuration of the object is initially unknown.

Due to considerable computational costs of ab initio method, the empirical data are often applied together with the apparatus of quantum mechanics. The application of experimental data and hypotheses in ab initio method resulted in the emergence of the family of semiempirical methods. The main hypothesis of semiempirical methods is that it is not electron orbitals that influence the interaction of particles but only valence ones. The problem is also simplified by substituting some integrals from the Schrödinger equation for algebraic and numerical analogs with the participation of experimental coefficients.

The accuracy of semiempirical methods is mainly determined by the selection of satisfactory empiric coefficients and adequate problem setting. Due to the decreased volume of calculations, these methods allow modeling more extended systems in comparison with ab initio methods.[5–8]

6.1.2 *Method of Molecular Dynamics*

The problem of atom deposition onto the template of porous aluminum oxide was solved with the method of molecular dynamics (MD). MD method is widely spread for modeling nanosystem behavior due to realization simplicity and high modeling accuracy. The system modeled in MD method is described as the aggregation of individual particles: molecules, atoms, and ions. Each molecule is split into separate atoms possessing certain properties and bonds with each other. MD method is based on the solution of Newton differential motion equation for each particle. For the system consisting of N atoms, the system of N vector equations with initial data is formed. Equations in the system are independent but require simultaneous solution[9–11]

$$m_i \frac{d^2\overline{\mathbf{r}}_i(t)}{dt^2} = \overline{\mathbf{F}}_i(t, \overline{\mathbf{r}}(t)), \quad i = 1, 2, .., N, \tag{6.10}$$

$$t_0 = 0, \overline{\mathbf{r}}_i(t_0) = \overline{\mathbf{r}}_{i0}, \frac{d\overline{\mathbf{r}}_i(t_0)}{dt} = \overline{\mathbf{V}}_i(t_0) = \overline{\mathbf{V}}_{i0}, \quad i = 1, 2, .., N \tag{6.11}$$

where N = number of atoms constituting the nanosystem;

m_i = mass of i atom;

$\overline{\mathbf{r}}_{i0}, \overline{\mathbf{r}}_i(t)$ = initial and current radius–vector of i atom, respectively, moreover, the atoms should not move outside the computational domain;

$\overline{\mathbf{F}}_i(t, \overline{\mathbf{r}}(t))$ = total force affecting i atom from the side of other atoms, $\overline{\mathbf{r}}(t)$ demonstrates the force dependence not only on the position of i atom, but also on the position of other atoms of the system;

$\overline{\mathbf{V}}_{i0}$, = initial velocity of i atom;

$\overline{\mathbf{V}}_i(t)$ = current velocity of i atom.

The function $\overline{\mathbf{F}}_i(t, \overline{\mathbf{r}}(t))$ in the eq 6.10 is represented as the gradient of potential energy:

$$\overline{\mathbf{F}}_i(t, \overline{\mathbf{r}}(t)) = -\frac{\partial U(\overline{\mathbf{r}}(t))}{\partial \overline{\mathbf{r}}_i(t)} + \overline{\mathbf{F}}_{ex}, \quad i = 1, 2, ..., N, \tag{6.12}$$

where $U(\overline{\mathbf{r}}(t))$ = potential energy value;

$\overline{\mathbf{r}}(t) = \{\overline{\mathbf{r}}_1(t), \overline{\mathbf{r}}_2(t), .., \overline{\mathbf{r}}_N(t)\}$ = demonstrates that potential energy depends on mutual position of all atoms;

$\overline{\mathbf{F}}_{ex}$ = external force acting upon i atom.

Depending on potential type and external forces in the eq 6.12, the problem of modeling the processes of nanofilm formation on porous templates of aluminum oxide will have different accuracy and different thermodynamic parameters. The issue of obtaining and searching the potential parameter is complicated and labor-consuming. The calculation with ab initio methods and experimental data will serve as the data sources. At present, the following experiments that allow obtaining the required parameters are known:

- atomic microscopes of high resolution provide the information about the valence structure parameters;
- spectroscopic data are the sources of oscillation frequencies and constants of deformations of bonds and valence angles;
- radiofrequency spectra allow obtaining the values of torsion oscillations and barriers of internal rotation;
- electric polarizabilities of molecules serve for finding Van der Waals potentials, dipole, and quadrupole moments of bonds of molecules and charges in atoms;
- thermophysical data are used for finding parameters of molecular liquids and energy of non-valence interatomic interactions.

The sets of quite properly chosen parameters for similar molecules are combined into special databases and libraries—force fields. When investigating the processes of obtaining nanofilms on porous templates of aluminum oxide with the help of molecular dynamics, the potentials and force fields applied are: embedded-atom method (EAM); Stillinger–Weber potential; modified embedded-atom method (MEAM); Abell–Tersoff potential; and force fields AMBER, CHARMM, GROMOS, OPLS, CVFF, and Merck Molecular Force Field (MMFF).[12–16]

The force field Assisted Model Building with Energy Refinement[15–16] (AMBER) is used for modeling the behavior of proteins, nucleic acids, and a number of other classes of molecules. It is not recommended to use this field to calculate properties of materials. Force fields built based on the calculations of molecular systems with the help of quantum-mechanical methods are available. MMFF can serve as example of such field and which is detailed in Ref. [17]. The force field Consistent Valence Force Field (CVFF) includes the refining contributions of anharmonicity and interaction of the force field components. The field GROningen MOlecular Simulation[8,19] (GROMOS) was developed within software package GROMACS and it comprises the parameters for wider family of chemical elements and their compounds.

6.1.3 Equations of Nanostructure Movement in Mesic Media

Due to small-time increment in MD method, condensation models can be used for modeling the formation of nanofilm coatings. The method of nanostructure movement in mesic media can be an example of such model. In this method, the movement and interaction of the condensed structures is considered. The solution of the system of differential equations for the interacting nanoparticles is the basis of this method:

$$m_i \frac{d^2 \bar{\mathbf{r}}_i(t)}{dt^2} = -m_i g + \bar{\mathbf{f}}_i(t) - m_i b_i \frac{d \bar{\mathbf{r}}_i(t)}{dt}, \; i = 1,2,...,n, \qquad (6.13)$$

where $\bar{\mathbf{f}}_i(t)$ = random force acting upon i nanoparticle;
b_i = "friction" coefficient;
n = number of nanoparticles.

The force $\bar{\mathbf{f}}_i(t)$ is similar to the random force in Langevin dynamics. $\bar{\mathbf{f}}_i(t)$ is found from Gaussian distribution with the following properties. The average value of the random force $\bar{\mathbf{f}}_i(t)$ equals zero. It is also assumed that it is not correlated with the velocity $\bar{\mathbf{V}}_i(t)$ of the nanoparticle examined, therefore, $<\bar{\mathbf{f}}_i(t)\bar{\mathbf{V}}_i(t)>$ equals zero and $<\bar{\mathbf{f}}_i(t)\bar{\mathbf{f}}_i(0)> = 2k_B T_0 b_i m_i \delta(t)$. Here, k_B is Boltzmann constant, $\delta(t)$ is Dirac delta function, and T_0 is initial temperature of the system.

To model the random force $\bar{\mathbf{f}}_i(t)$, Box–Mueller transformation was used in the eq 6.13. The method is accurate, contrary to, for example, the methods based on the central limit theorem.

Suppose x and y are independent random values, uniformly distributed on the interval $[-1, 1]$. $R = x^2 + y^2$ is found. If $R > 1$ or $R = 0$, the values of x and y are generated one more time. As soon as the assumption $0 < R \leq 1$ is fulfilled, z_0 and z_1 are calculated by the eqs 6.14 and 6.15, which will be independent random values satisfying the standard normal distribution

$$z_0 = x \cdot \sqrt{\frac{-2 \ln R}{R}}, \qquad (6.14)$$

$$z_1 = y \cdot \sqrt{\frac{-2 \ln R}{R}} \qquad (6.15)$$

After the standard normal random value z is obtained, the transition to the value $\xi \sim N\left(\mu, \sigma^2\right)$ is performed, normally distributed with the mathematical expectation μ and standard deviation σ, by the formula

$$\xi = \mu + \sigma \cdot z \tag{6.16}$$

In accordance with the data on the random force $\overline{\mathbf{f}}_i(t)$ in the eq 6.13, the force has the mathematical expectation $\mu = 0$ and standard deviation $\sigma = \sqrt{2k_B T_0 b_i m_i \delta(t)}$.

The condensation of atoms, molecules, ions, and nanostructures in the method of molecular dynamics is conditioned by the availability of certain potentials between atoms. Two main factors influence the condensation of nanoparticles in mesic media:

- distance between interacting particles;
- direction and value of velocities.

Two different nanoparticles are at the distance R_{ij} in the arbitrary point of time (Fig. 6.1). The selection of "agglutination" assumption is substantiated rather properly, when R_{ij} is small, that is, in the event of contact between the nanoparticles.

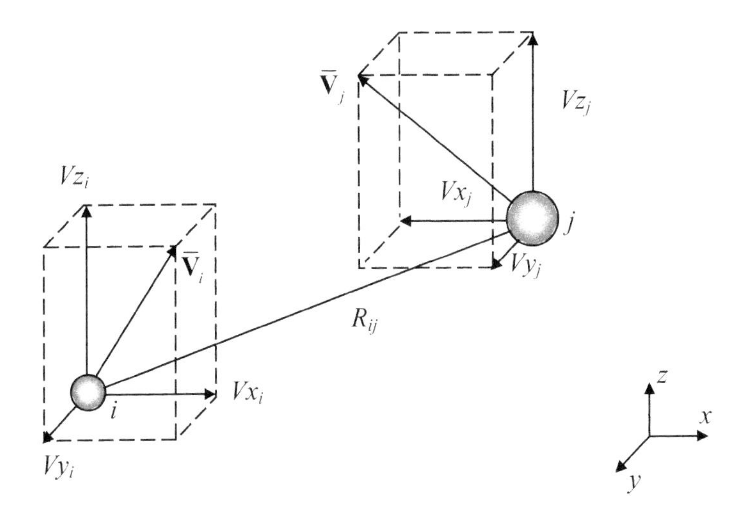

FIGURE 6.1 Mutual arrangement of i and j nanoparticles.

The second factor influencing the condensation of nanostructures, is found by the values and direction of velocities. It is obvious that very "fast" nanoparticles can "slip" past each other even with the contact. The angle α between the velocity vectors also significantly influences the condensation of nanoparticles (Fig. 6.2), which determines the movement direction. The selection of adequate condition of nanostructure condensation is important, as it determines the legitimacy of transition from the stage of molecular dynamics to the stage of nanostructure movement in mesic media.

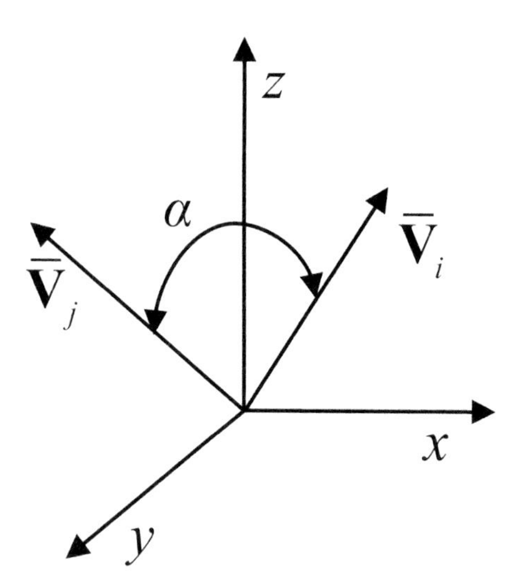

FIGURE 6.2 Mutual arrangement of the velocities of *i* and *j* nanostructures.

To investigate the processes of movement and condensation of nanoparticles with the help of the movement method in mesic media, only the grouped nanoparticles were considered and the influence of the molecules of gas phase was substituted for the influence of the random force. The transition from molecular dynamics to the method of particle movement allowed increasing the spatial size of the computational cell. Since the problem was solved using the periodic boundary conditions, the scale by space was increased by the symmetric imaging of the grouped nanoparticles onto the neighboring computational cells. Besides, due to the increased mass of the nanoparticles investigated, it became possible to increase the integration step.

The calculated volume was increased several times. With sufficient consolidation of nanoparticles, their concentration in the calculated volume significantly decreased. There was the situation when the nanoparticles practically did not interact within the calculated volume. However, if we take into account the influence of nanoparticles from the neighboring cells, the condensation process continues that indicates the necessity to consider the problem solved in larger spatial scale. Thus, for adequate investigation of the condensation problem, it is necessary to timely increase the spatial scale of the cell and merging several calculated volumes into one (Fig. 6.3).

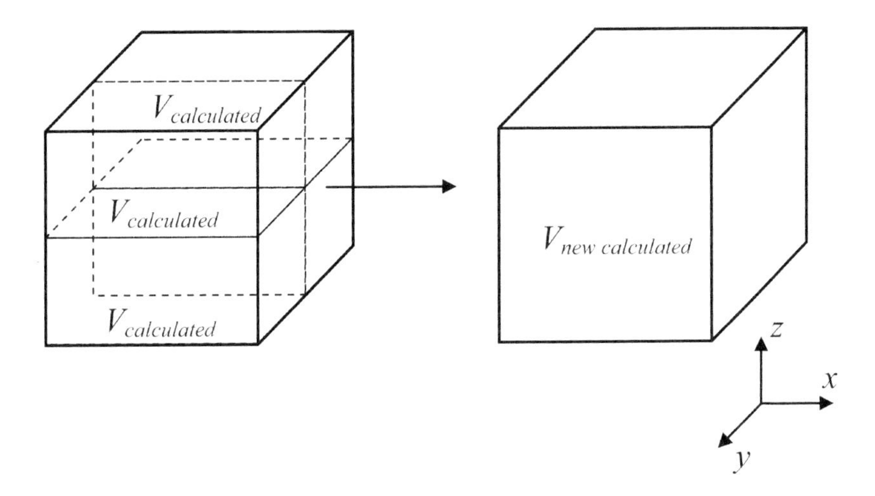

FIGURE 6.3 Calculated volume increase.

The search for the adequate integration step plays an important role in the problem of nanostructure condensation with the help of the particle movement method. The selection of the small step by time increases the time for the problem calculation. Even at the current stage of computer engineering development, several dozen years can be required to solve the problem of molecular dynamics to real-time scales. The excessive integration step also results in undesirable results: the situation arises when nanoparticles will "slip" past each other. Thus, the problem of selecting the optimum step by time emerges.

The solution of this issue in the paper is based on the values of nanoparticle coordinates and velocity values. The arbitrary pair of particles at some time moment in shown in Figure 6.4. The particle velocities are $\bar{\mathbf{V}}_i$ and $\bar{\mathbf{V}}_j$, the position is determined by $\bar{\mathbf{r}}_i$ and $\bar{\mathbf{r}}_j$—current radius-vectors for i and j nanoparticles, respectively.

The projection of velocities of i and j nanostructures are found by the formulas:

$$pr_{R_{ij}}\bar{\mathbf{V}}_i = \bar{\mathbf{V}}_i \cdot \bar{\mathbf{R}}_{ij} / R_{ij}, \tag{6.17}$$

$$pr_{R_{ij}}\bar{\mathbf{V}}_j = \bar{\mathbf{V}}_j \cdot \bar{\mathbf{R}}_{ji} / R_{ij} \tag{6.18}$$

The vectors $\bar{\mathbf{R}}_{ij}, \bar{\mathbf{R}}_{ji}$ and distance between the nanostructures R_{ij} are calculated from the formulas:

$$\bar{\mathbf{R}}_{ij} = \bar{\mathbf{r}}_j - \bar{\mathbf{r}}_i, \bar{\mathbf{R}}_{ji} = \bar{\mathbf{r}}_i - \bar{\mathbf{r}}_j, \tag{6.19}$$

$$R_{ij} = \left|\bar{\mathbf{R}}_{ij}\right| = \left|\bar{\mathbf{R}}_{ji}\right| = \left|\bar{\mathbf{r}}_j - \bar{\mathbf{r}}_i\right| \tag{6.20}$$

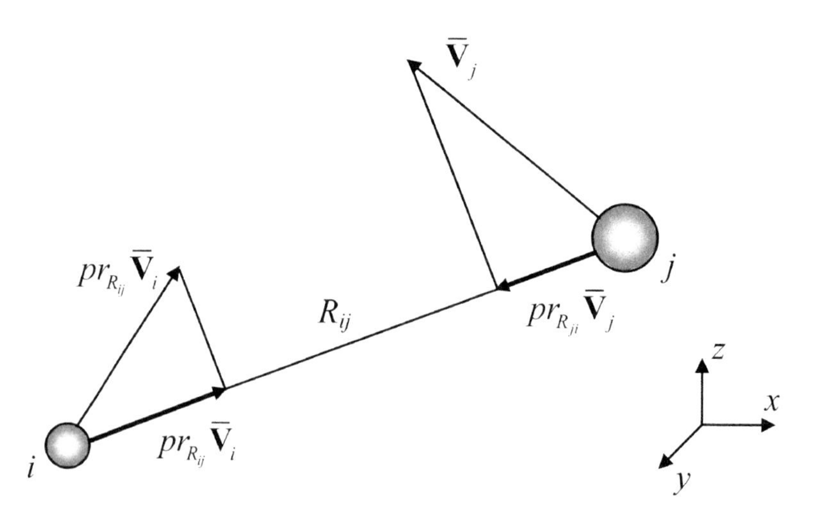

FIGURE 6.4 Search for the integration step.

The time interval after which the collision of i and j nanoparticles becomes possible is directly proportional to the distance between nanoparticles and inversely proportional to the velocity projections:

$$\Delta t_{ij} = \frac{R_{ij}}{pr_{R_{ij}} \bar{\mathbf{V}}_i + pr_{R_{ij}} \bar{\mathbf{V}}_j} \tag{6.21}$$

By the eq 6.21, the least positive value of Δt_{ij} is found and the integration step is calculated taking into consideration all possible nanoparticle pairs:

$$\Delta t = \frac{1}{m} \cdot \min_{i,j,\,dt>0} \left(\Delta t_{ij}\right); i = 1,2,...,n; j = i, i+1,...,n, \tag{6.22}$$

where m is integer number determines what part of the period of the fastest particles is the step by time. The integration step selection is influenced, first of all, by the movement of the fastest particles. In order not to delay the problem solution with additional calculations, it is rational to select the integration step not at every iteration but after certain number of steps by time.

6.2 POTENTIALS OF FORCE FIELDS

The type of potentials and potential energy of nanostructured systems and nanofilm coatings contribute decisively to the type, character, and value of interactions of nanosystem objects. Potentials are divided into multiparticle and paired, axisymmetric and spatial. When solving the problem, the potential type is mainly determined by the availability of parameters in libraries and force fields of databases for modeling nanostructured systems.

6.2.1 *Intermolecular Potential*

It is used when the nanostructured system is formed by the molecules and some atoms or ions are limited by bonds. In this case, the potential energy of interacting atoms is represented as the total of contributions from different types of interactions between the atoms[20,21]:

$$U(\overline{\mathbf{r}}(t)) = U_b + U_v + U_\varphi + U_{vdw} + U_q, \tag{6.23}$$

where the summands correspond to the following types of atom interactions in the molecule:

U_b = chemical bonds;
U_v = valence angles;
U_φ = torsion angles and plane groups;
U_{vdw} = Van der Waals forces;
U_q = electrostatic interactions.

Its own law is introduced for each interaction type, which comprises the parameters of atoms defining the atom properties and behavior in the system. As the distribution of atom electron clouds in the molecule is individual, the parameters of atom interaction for different types of molecules will be different.

The valence lengths of chemical bonds in eq 6.23 shown in Figure 6.5 are maintained due to the potential:

$$U_b = \sum_{bonds} K_b (b - b_0)^2, \tag{6.24}$$

where K_b = constant of the bond elongation-compression;
b_0 = equilibrium bond length;
b = current bond length;
bonds = number of bonds in the molecule.

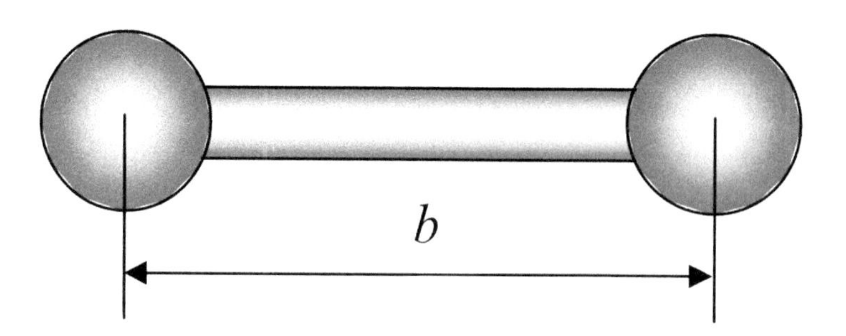

FIGURE 6.5 Change in the distance between two atoms, where b = bond value.

The potential of valence angles of the molecule is determined by the formula, the interaction is depicted in Figure 6.6:

$$U_v = \sum_{angles} K_\theta(\theta_{an} - \theta_{an0})^2, \tag{6.25}$$

where K_θ = bend force constant;

θ_{an0} = equilibrium value of the bond angle;

θ_{an} = current value of the bond angle;

the summing up parameter *angles* indicates that the summing up is carried out along all the molecule angles.

The energy value of torsion interactions and plane groups is written down as the expression, the scheme is given in Figure 6.7:

$$U_\varphi = \sum_{dehidral} \begin{cases} K_\varphi(1+\cos(l\varphi-\delta)), l \neq 0 \\ K_\varphi(\varphi-\delta)^2, l = 0, \end{cases} \tag{6.26}$$

where l = multiplicity factor of the torsion barrier;

δ = phase shift;

constant K_φ = characterizes the height of potential barriers of dihedral angles φ;

dehidral = number of dihedral angles.

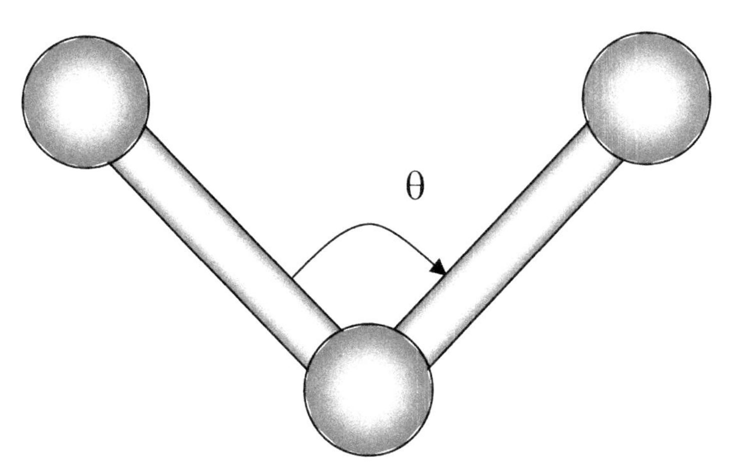

FIGURE 6.6 Change in the angle between three atoms, where θ = angle value.

The energy of plane groups is described via the so-called “irregular” dihedral angles and calculated with the help of the eq 6.26.

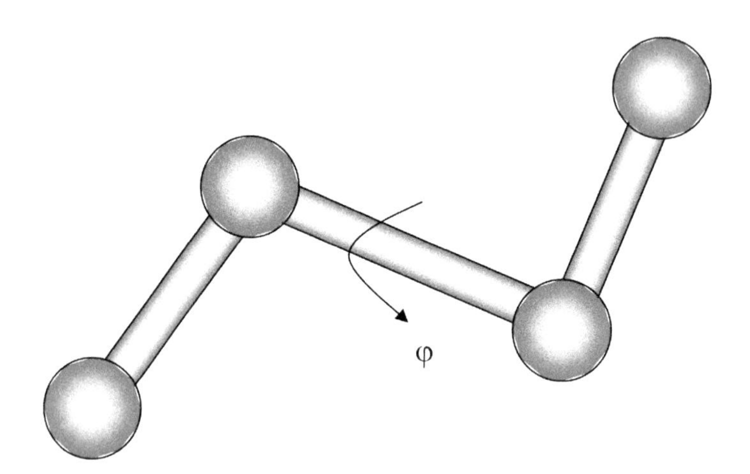

FIGURE 6.7 Change in the dihedral angle between four atoms, where φ = dihedral angle value.

Van der Waals interactions are given as Lennard-Jones potential:

$$U_{vdw} = \sum_{ij \in vdw} \left[\frac{A_{ij}}{R_{ij}^{12}} - \frac{B_{ij}}{R_{ij}^{6}} \right], \tag{6.27}$$

where A_{ij} and B_{ij} = interaction parameters.

$$A_{ij} = \left(\frac{e_i^*}{2} + \frac{e_j^*}{2} \right)^{12} \sqrt{\varepsilon_i \varepsilon_j}\ ;\ B_{ij} = 2\left(\frac{e_i^*}{2} + \frac{e_j^*}{2} \right)^{6} \sqrt{\varepsilon_i \varepsilon_j}, \tag{6.28}$$

where $\frac{e_i^*}{2}$ = half of the distance at which the separation energy of two atoms of i type is minimum;

$\frac{e_j^*}{2}$ = half of the distance at which the separation energy of two atoms of j type is minimum;

ε_i = depth of the potential well of the energy of Van der Waals interaction for i atom;

ε_j = depth of the potential well of the energy of Van der Waals interaction for j atom;

R_{ij} = distance at which the interaction takes place.

Electrostatic interactions are calculated by the interaction formula for two point charges:

$$U_q = \sum_j \frac{q_i q_j}{\varepsilon R_{ij}}, \tag{6.29}$$

where q_i, q_j = partial charge on atoms;

ε = dielectric penetration of the medium;

R_{ij} = distance between the pair of atoms, at which the electrostatic interaction occurs.

When taking into account the paired interatomic interaction during the mathematical modeling of metal and/or semiconductor systems, several problems come out. In Ref. [22], it was indicated as follows:

(1) when using only paired interaction potential, the non-physical correlation for Cauchy coefficients in the systems "metal and/or semiconductor" is fulfilled ($C_{12} = C_{44}$);
(2) when using only multi-particle interaction potentials, the following correlations are fulfilled: $C_{12} = C_{11}$, $C_{44} = 0$. This contradicts the experiments.

The paired potentials cannot provide the realistic values of physical characteristics of materials.[23] For correct description of solid properties, it is necessary to use multi-particle potentials. It is known that none of the existing potentials can reproduce the complete set of characteristics of solids.

Thus, the potential selection for mathematical modeling is a complicated complex problem. The majority of empiric potentials describe the material volumetric properties well, but, nevertheless, some of them are also successfully used to describe surface properties.

The actual correlations of elastic constants of metals and semiconductors can be obtained by only taking into account paired and multi-particle interactions. The following approaches considering multi-particle interaction are most widely spread when modeling metal and semiconductor systems:

- Stillinger–Weber potential[24];
- Abel–Tersoff potential[25,26];
- embedded atom method[22,27,28];
- modified embedded atom method.[29]

6.2.2 Lennard-Jones Potential

Lennard-Jones potential[30,31] is widely applied to calculate properties of gases, liquids, and solids, in which the intermolecular forces occur. The potential describes the paired interaction of spherical non-polar molecules and comprises the dependence of interaction energy of two particles on the distance between them—repulsion at small distances and attraction—at the large ones. Lennard-Jones potential is as follows:

$$U^{LD}(r) = 4\varepsilon\left(\left(\frac{\sigma}{r}\right)^{12} - \left(\frac{\sigma}{r}\right)^{6}\right), \tag{6.30}$$

where ε = depth of the potential well;
r = distance between the particle centers;
σ = distance at which the interaction energy becomes zero.

The values ε and σ are determined by the material properties. The characteristic form of the potential is demonstrated in Figure 6.8.

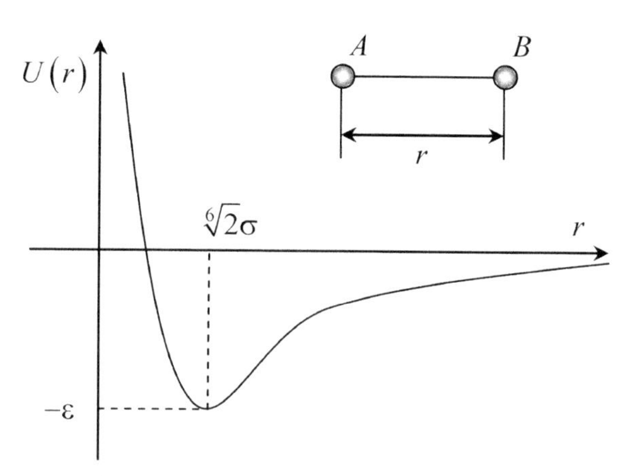

FIGURE 6.8 Lennard-Jones potential.

This potential is attracting at large distances, has the minimum in the point $r_{min} = \sqrt[6]{2}\sigma$ and is repulsing at small distances. When $r > r_{min}$, the attraction forces prevail over the repulsion forces, thus corresponding to the summand of the formula $(\sigma/r)^6$. The nature of attraction forces is explained by dipole–dipole induced interaction[32] (Van der Waals forces). When the distance between the particles $r < r_{min}$, the repulsion forces prevail over the attraction forces due to the exchange interaction that corresponds to the component $(\sigma/r)^{12}$ of the formula (during the overlapping of electron clouds the molecules start to repulse actively). The shortcoming of such representation of interaction potential is the fact that the forces of exchange interaction responsible for particle interaction at short distances are only approximately described by the power dependence. Physically, it is more correct to select the repulsion part in exponential form. However, the potential power representation is convenient for computer calculations. For this reason, Lennard-Jones potential is widely used in numerical modeling of substance behavior.

Lennard-Jones potential is also frequently written down as follows:

$$U^{LD}(r) = \varepsilon\left(\left(\frac{r_{min}}{r}\right)^{12} - 2\left(\frac{r_{min}}{r}\right)^{6}\right), \tag{6.31}$$

where $r_{min} = \sqrt[6]{2}\sigma$ is potential minimum point.

To accelerate the calculations, Lennard-Jones potential is usually cut at the distance $r_c = 2{,}5\sigma - 3{,}5\sigma$. This selection of $r_c = 2{,}5\sigma$ is conditioned by the fact that at this distance, the interaction energy value is only 1.63 % of the well depth ε.[33] To avoid the sharp change in the potential, and, consequently, the system energy kick in case of the potential cut down, it is gradually decreased to zero.

Lennard-Jones potential is a two-parameter one, therefore, it does not fit other molecule types (non-spherical or having constant dipole moments). On the other hand, this potential quite accurately describes the properties of a number of substances (e.g., crystalline inert gases) and Van der Waals interaction forces. The advantages of Lennard-Jones potential also comprise its computational simplicity, which does not require the calculation of irrational and transcendental functions. Lennard-Jones potential is applied as a classical model potential, whose main task is to describe general physical regularities and not to obtain accurate numerical results.[34,35]

6.2.3 Stockmayer Potential

Stockmayer potential is a simple model of the paired interaction of molecules with constant dipole moment. This potential is the modified Lennard-Jones potential with additional component of dipole interaction.[36,37] For the molecule pair with constant dipole moments, the energy of their interaction will depend not only on the distance between them but also on their mutual orientation. Stockmayer potential is described by the following formula:

$$U^{St}(r,\theta_A,\theta_B,\varphi)=4\varepsilon\left(\left(\frac{\sigma}{r}\right)^{12}-\left(\frac{\sigma}{r}\right)^{6}\right)-\frac{\mu_A\mu_B}{r^3}g(\theta_A,\theta_B,\varphi), \quad (6.32)$$

where r = distance between the molecules;
σ,ε = parameters of Lennard-Jones potential;
θ_A,θ_B = polar angles of the molecules;
$\varphi=\varphi_B-\varphi_A$ = difference of azimuth angles of the molecules;
μ_A,μ_B = diploe moments of the molecules, the function g is calculated by the formula $g(\theta_A,\theta_B,\varphi)=2\cos\theta_A\cos\theta_B-\sin\theta_A\sin\theta_B\cos\varphi$.

The meaning of these values is explained in Figure 6.9.

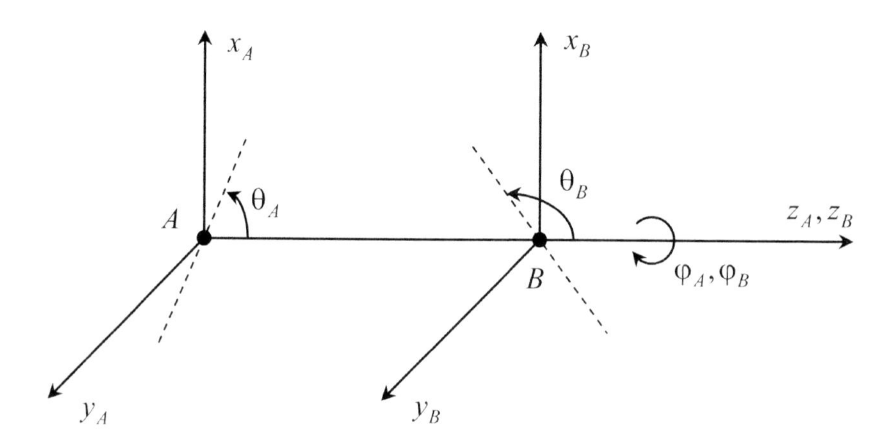

FIGURE 6.9 Parameters of Stockmayer potential.

Stockmayer potential is presented as suspension of two simpler potentials $U^{St} = U^{LD} + U^{d}$. The first is Lennard-Jones potential, the second is interaction potential of two diploes. There is an alternative recording form of Stockmayer potential. If the dimensionless dipole moment $\mu^* = \mu/\sqrt{\varepsilon\sigma^3}$ is introduced, the potential is as follows:

$$U^{St}(r,\theta_A,\theta_B,\varphi) = \varepsilon\left[4\left(\left(\frac{\sigma}{r}\right)^{12} - \left(\frac{\sigma}{r}\right)^{6}\right) - \mu^{*2} g(\theta_A,\theta_B,\varphi)\left(\frac{\sigma}{r}\right)^{3}\right] \quad (6.33)$$

Stockmayer potential inherits all the limitations of Lennard-Jones potential. Besides, Stockmayer potential uses expressions for the interaction of point dipoles but real molecules have finite dimensions.

6.2.4 Morse Potential

The nuclei potential energy in lowest electron states of real diatomic molecules can be quite accurately described with the help of simple analytical function called "Morse potential." Morse potential[38,39] is paired force interaction potential and can be represented as follows:

$$U^{Morse}(r) = \varepsilon\left(e^{-2\alpha(r-\sigma)} - 2e^{-\alpha(r-\sigma)}\right), \quad (6.34)$$

where ε = bond energy;
σ = equilibrium interatomic distance;
α = parameter characterizing the width of the potential well.

Morse potential is a three-parameter one and is spread due to its simplicity and visual expression. Its advantage comprise fast attenuation at the distance, if during modeling it is necessary to take into account only the interaction of the nearest particles.

In the dimensionless form, if we take $\alpha = \sigma$, the potential dependence is represented by the following eq 6.35, whose graphical form is demonstrated in Figure 6.10.

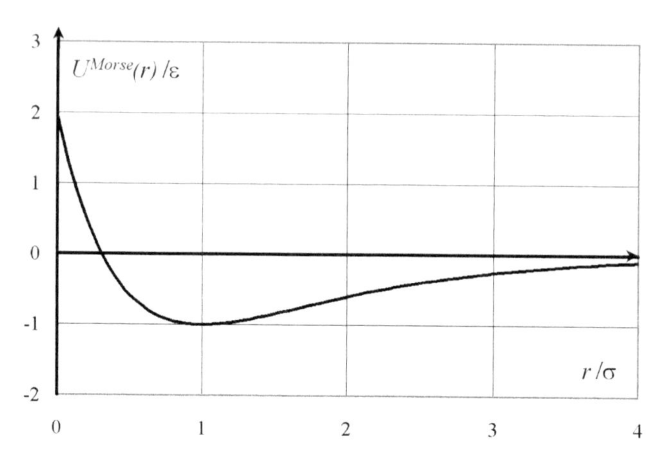

FIGURE 6.10 Morse potential at $\alpha=\sigma$.

$$\frac{U^{Morse}(r)}{\varepsilon}=\left(e^{-\frac{2\alpha}{\sigma}\left(\frac{r}{\sigma}-1\right)}-2e^{-\frac{\alpha}{\sigma}\left(\frac{r}{\sigma}-1\right)}\right)=e^{-2\left(\frac{r}{\sigma}-1\right)}-2e^{-\left(\frac{r}{\sigma}-1\right)} \quad (6.35)$$

The function of Morse potential energy reaches its minimum $U^{Morse}(r)=-\varepsilon$ at $r=\sigma$, and exponentially tends to zero at large r. Morse potential energy is built based on the theoretical investigations of short-range part of the potential. The energy value corresponding to neutral atoms separated apart is arbitrary taken as "zero."

6.2.5 *Mi Potential*

Mi potential is another paired force potential.[40,41] It is determined by the following formula:

$$U^{Mi}(r)=\frac{\varepsilon}{n-m}\left(m\left(\frac{\sigma}{r}\right)^{n}-n\left(\frac{\sigma}{r}\right)^{m}\right), \quad (6.36)$$

where ε = bond energy;

σ = bond length;

m and n = dimensionless parameters determining rigidity and long-range action of the paired potential.

In particular case, when $m=6, n=12$, Mi potential transforms into Lennard-Jones potential.

6.2.6 Buckingham Potential

Buckingham potential[42,43] is applied to describe Van der Waals interactions and is based on the assumption of exponential dependence of repulsion forces between molecules on the distance between them. In this case, the potential energy of interaction between the molecules is described by the following formula:

$$U^{Buck}(r) = ae^{-br} - \frac{c}{r^6} - \frac{d}{r^8}, \tag{6.37}$$

where a, b, c, d are empirical parameters, which are selected by approximation to true experimental data.

Buckingham potential is the combination of repulsion potential in exponential form with the addition of binding components of dispersion contribution $-c/r^6$ and $-d/r^8$. The eq 6.37 is correct only for nonpolar spherically symmetric molecules. The corrections are introduced into the empirical potential to calculate the interaction energy of more complicated systems.[44]

The systems are frequently modeled with the help of simplified Buckingham potential, in which the third summand $-d/r^8$ is neglected and the whole interaction is considered as the total of interactions between all atoms of the subsystems. The analysis of the equation approximation (eq 6.37) demonstrates that it is possible to quite adequately use Buckingham formula without the third component when finding structural and physical-mechanical properties of disperse systems.[45] It is considered that Buckingham potential more accurately describes the interaction character of atom pairs at distances corresponding to typical lengths of chemical bonds, but Lennard-Jones potential is applied at any interatomic distances.

6.2.7 Stillinger–Weber Potential

Stillinger–Weber potential[46] is one of the first potentials describing materials forming diamond-like structure (C, Si, GaAs, Ge). The potential energy of this potential is written down as follows:

$$E = \sum_{i}\sum_{j>i}\phi_2(r_{ij}) + \sum_{i}\sum_{j\neq i}\sum_{k>j}\phi_3(r_{ij}, r_{ik}, \theta_{ijk}), \tag{6.38}$$

$$\phi_2\left(r_{ij}\right)=A_{ij}\varepsilon_{ij}\left[B_{ij}\left(\frac{\sigma_{ij}}{r_{ij}}\right)^{p_{ij}}-\left(\frac{\sigma_{ij}}{r_{ij}}\right)^{q_{ij}}\right]\exp\left(\frac{\sigma_{ij}}{r_{ij}-a_{ij}\sigma_{ij}}\right), \tag{6.39}$$

$$\phi_3\left(r_{ij},r_{ik},\theta_{ijk}\right)=\lambda_{ijk}\varepsilon_{ijk}\left[\cos\theta_{ijk}-\cos\theta_{0ijk}\right]^2\exp\left(\frac{\gamma_{ij}\sigma_{ij}}{r_{ij}-a_{ij}\sigma_{ij}}\right)\exp\left(\frac{\gamma_{ik}\sigma_{ik}}{r_{ik}-a_{ik}\sigma_{ik}}\right), \tag{6.40}$$

where A, B, p, q, λ, γ, σ, ε are empirical parameters;
θ is tetrahedral angle.
The main problem of this potential is its bad transmission.

6.2.8 Abell–Tersoff Potential

The general form of this potential was proposed by Abell,[47] after that it was modified by Tersoff.[48] This potential describes well the mechanical properties, elastic properties, and dependence of the dynamics of these structures on temperatures.

Tersoff potential depends not only on the distances between the atoms, but also on the angle between three atoms. This makes the potential more flexible and accurate in computations. The potential contains the information about the angle θ between two atoms. Without this angle, the number of possible atom positions is unlimited and does not correlate with physical reality. The energy function is written down as follows:

$$E=\frac{1}{2}\sum_i\sum_{j\neq i}V_{ij}, \tag{6.41}$$

where V_{ij} = bond energy between i and j atoms:

$$V_{ij}=f_C\left[f_B\left(r_{ij}\right)+b_{ij}f_A\left(r_{ij}\right)\right] \tag{6.42}$$

The functions f_A and f_B are responsible for the attraction and repulsion of atoms, respectively. The function f_c is responsible for the bond length extreme and cutting smoothness. The parameters R and D are selected in such a way as to limit the interactions between the first nearest neighbors:

$$f_A(r) = -Be^{-\lambda_2 r}, \tag{6.43}$$

$$f_B(r) = Ae^{-\lambda_1 r}, \tag{6.44}$$

$$f_C(r) = \begin{cases} 1, & r < R - D \\ \frac{1}{2} - \frac{1}{2}\sin\frac{\pi(r-R)}{2D}, & R - D < r < R + D \\ 0, & r > R + D \end{cases} \tag{6.45}$$

$$b_{ij} = \left(1 + \beta^n \xi_{ij}^n\right)^{-1/2n}, \tag{6.46}$$

$$\xi_{ij} = \sum_{k \neq (i,j)} f_C(r_{ik}) g(\theta_{ijk}) e^{\lambda_3^3 (r_{ij} - r_{ik})} \tag{6.47}$$

The parameter b_{ij} is responsible for the bond length. The bond length depends on the local position of the atom. In this case, the more neighbors of an atom, the shorter thelengthof atom bonds. Tersoff potential is based on Morse potential.

6.2.9 *Embedded Atom Method*

In embedded atom method (EAM), the bond energy of atoms is given as follows:

$$E = \sum_i F_i\left[\sum_{j \neq i} \rho_j(R_{ij})\right] + \frac{1}{2}\sum_{j \neq i} \varphi_{ij}(R_{ij}), \tag{6.48}$$

where $\sum_i F_i\left[\sum_{j \neq i} \rho_j(R_{ij})\right]$ is function of atom i embedding, depending on the total electron density in the region of i embedding;

$\varphi_{ij}(R_{ij})$ = energy of paired interaction.

The conclusion (eq 6.48) with the use of density functional theory (DFT) can be found in Ref. [28].

Each atom of the system is considered as a particle embedded into the electron gas produced by the rest of the atoms of the system under consideration. The energy required for embedding depends on the electron density in the embedding point. The embedding function developed so, allows finding the exchange and correlation energies of the system electron gas.

The sense of the embedding function can be defined as the energy necessary for embedding one atom into the homogeneous electron gas with the density ρ. However, there are other transformations as well[49] that allow changing the functions (eq 6.48) on the condition that the resultant energy and interatomic forces will not change.

The following approximations are used in:

(1) function of electron density of one atom is a spherically symmetrical function depending only on the distance between atoms. This approximation significantly limits the area of EAM application and allows considering the systems, in which the directivity of the bond covalence component can be neglected;

(2) electron density in the embedding region of atom i is found as the linear superposition of electron densities of the rest of the system atoms $\sum_{j \neq i} \rho_j(R_{ij})$. This approximation significantly simplifies the electron density computation;

(3) value $\sum_{j \neq i} \rho_j(R_{ij})$ in metal systems in the region of i atom location changes slightly in comparison with the electron density of atom ρ_i itself. Thus, the total density $\sum_{j \neq i} \rho_j(R_{ij})$ in the region of i atom location can be replaced by constant $\bar{\rho}$.[28] The energy of electron gas density is approximated by the function depending only on the mean value of the electron density in the embedding region and not by complex functionals as in DFT method.

At present, EAM potentials have been developed for the majority of metals and some binary systems. The potentials for ternary systems were also calculated.[50] However, such "ternary" potentials do not qualitatively reproduce physical properties of materials.

6.2.10 Modified Embedded-Atom Method

The semiempirical approach combining the advantages of multi-particle potentials and embedded-atom method was proposed in 1990s. The theory of modified embedded-atom method (MEAM) is developed with the application of electron functional theory (DFT).[51] At present, DFT method is considered as the most acknowledged approach to describe the electron properties of solids. In EAM method the electron density is represented as linear superposition of spherically averaged functions. This disadvantage is eliminated in MEAM.

In MEAM, the system bond energy is written down as follows:

$$E = \sum_i \left(F_i \left(\frac{\bar{\rho}_i}{Z_i} \right) + \frac{1}{2} \sum_{j \neq i} \varphi_{ij} \left(R_{ij} \right) \right), \tag{6.49}$$

where E = energy of i atom;

F_i = embedding function for atom i embedded into electron density $\bar{\rho}_i$;

Z_i =– number of nearest neighbors of atom i in its reference crystalline structure;

φ_{ij} =– paired potential between atoms i, j located at distance R_{ij}.

In MEAM, the embedding function $F(\rho)$ is found as follows:

$$F(\rho) = A_p E_c \rho \ln \rho, \tag{6.50}$$

where A_p = controllable parameter;

E_c = bond energy.

The paired potential between the atoms i, j is found by the following formula:

$$\varphi_{ij}(R) = \frac{2}{Z_i} \left\{ E_i^u(R) - F_i \left(\frac{\bar{\rho}_i^0(R)}{Z_i} \right) \right\}, \tag{6.51}$$

where $\bar{\rho}_i^0$ = electron density.

The electron density in the embedding point comprises angular dependencies and is written down as follows:

$$\bar{\rho} = \rho^{(0)} G(\Gamma) \tag{6.52}$$

There are many varieties for function G[52]:

$$G(\Gamma) = \sqrt{1+\Gamma}, \tag{6.53}$$

$$G(\Gamma) = e^{\frac{\Gamma}{2}}, \tag{6.54}$$

$$G(\Gamma) = \frac{2}{1+e^{-\Gamma}}, \tag{6.55}$$

$$G(\Gamma) = \pm\sqrt{|1+\Gamma|} \tag{6.56}$$

The following record form is most widely used:

$$G(\Gamma) = \sqrt{1+\Gamma} \tag{6.57}$$

Function G is calculated by the following formula:

$$\Gamma = \sum_{h=1}^{3} t^{(h)} \left(\frac{\rho^{(h)}}{\rho^{(0)}} \right)^2, \tag{6.58}$$

where $h = 0-3$, correspond to *s, p, d, f* symmetries;
$t^{(h)}$ = weighting factors;
$\rho^{(h)}$ = values defining the deviation of electron density distribution from the distribution of cubic system in ideal crystal $\rho^{(0)}$:

$$s(h=0) : \rho^{(0)} = \sum_i \rho^{a(0)}(r^i), \tag{6.59}$$

$$p(h=1) : \left(\rho^{(1)}\right)^2 = \sum_\alpha \left[\sum_i \rho^{a(1)}(r^i) \frac{r_\alpha^i}{r^i} \right]^2, \tag{6.60}$$

$$d(h=2):\left(\rho^{(2)}\right)^2=\sum_{\alpha,\beta}\left[\sum_i \rho^{a(2)}\left(r^i\right)\frac{r_\alpha^i r_\beta^i}{r^{2i}}\right]^2-\frac{1}{3}\sum_i\left[\sum_i \rho^{a(2)}\left(r^i\right)\right]^2, \quad (6.61)$$

$$f(h=3):\left(\rho^{(3)}\right)^2=\sum_{\alpha,\beta,\gamma}\left[\sum_i \rho^{a(3)}\left(r^i\right)\frac{r_\alpha^i r_\beta^i r_\gamma^i}{r^{3i}}\right]^2, \quad (6.62)$$

where $\rho^{a(h)}$ = radial functions, which represent the distribution decrease by distance r^i, superscript i indicates the nearest atoms;

α,β,γ = summation indexes by each of the three possible directions.

Finally, the individual contribution is calculated by the following formula:

$$\rho^{a(h)}(r)=\rho_0 e^{-\beta^{(h)}\left(\frac{r}{r_e}-1\right)} \quad (6.63)$$

Different interaction potentials were applied when solving the problem of modeling the processes of obtaining special nanofilm coatings on the templates of porous aluminum oxide. The main attention was paid to Lennard-Jones potential, intermolecular potential, and MEAM.

6.3 NUMERICAL METHODS FOR SOLVING THE DEPOSITION PROBLEM

The eq 6.10 represents the system of ordinary differential equations of second order with initial conditions (eq 6.11). There are different ways of solving similar equations,[53] for example, integral-interpolational methods. One of the ways of solving equations is its transition to the system of differential equations of first order, with the help of substitution—the system first equation:

$$\begin{cases}\dfrac{d\overline{\mathbf{r}}(t)}{dt}=\overline{\mathbf{V}}(t),\\ m\dfrac{d\overline{\mathbf{V}}(t)}{dt}=\overline{\mathbf{F}}\left(t,\overline{\mathbf{r}}(t),\overline{\mathbf{V}}(t)\right).\end{cases} \quad (6.64)$$

Cauchy problem for the system (eq 6.64) can be solved by different numerical methods. Euler, Adams, Runge-Kutta methods, and predictor-corrector algorithms are used to solve differential equations of first order.[53,54] Due to the fact that the system equations are similar and take the form $u'(t) = f(t)$, only the second equation of the system will be further considered for the development of numerical algorithms. Let us assume that the solution $\bar{\mathbf{V}}(t)$ on the time interval t is known. The problem of finding the solution $\bar{\mathbf{V}}(t_1)$ on the new step by time $t_1 = t + \Delta t$ arises, where Δt is value of time step.

6.3.1 Euler Methods

Due to its implementation simplicity, Euler method holds the key position when solving differential equations of first order. The solution expansion $\bar{\mathbf{V}}(t+\Delta t)$ into Taylor series of k order in point t by degrees $t_1 - t = \Delta t$ is written down as follows:

$$\bar{\mathbf{V}}(t+\Delta t) = \bar{\mathbf{V}}(t) + \Delta t \frac{d\bar{\mathbf{V}}(t)}{dt} + \frac{1}{2!}\Delta t^2 \frac{d^2\bar{\mathbf{V}}(t)}{dt^2} + \ldots +$$

$$+ \frac{1}{k!}\Delta t^k \frac{d^k\bar{\mathbf{V}}(t)}{dt^k} + O\left(\Delta t^{k+1}\right) \tag{6.65}$$

The value $\bar{\mathbf{V}}(t_1) = \bar{\mathbf{V}}(t+\Delta t)$ is the desired solution and summand $O\left(\Delta t^{k+1}\right)$ indicates that the error during the solution expansion into Taylor series will be the value of $k+1$ order of small quantity in relation to time step Δt. If at the right side of the expansion (eq 6.65) we are restricted only by two first summands, we have the formula of ordinary Euler method:

$$\bar{\mathbf{V}}(t+\Delta t) = \bar{\mathbf{V}}(t) + \Delta t \frac{d\bar{\mathbf{V}}(t)}{dt} + O\left(\Delta t^2\right) \tag{6.66}$$

The second equation of the system (eq 6.66) gives the formula

$$\frac{d\bar{\mathbf{V}}(t)}{dt} = \frac{\bar{\mathbf{F}}\left(t, \bar{\mathbf{r}}(t), \bar{\mathbf{V}}(t)\right)}{m} \tag{6.67}$$

and allows turning into Euler method in the following form:

$$\bar{\mathbf{V}}(t+\Delta t)=\bar{\mathbf{V}}(t)+\Delta t\frac{\bar{\mathbf{F}}(t,\bar{\mathbf{r}}(t),\bar{\mathbf{V}}(t))}{m}+O(\Delta t^2) \quad (6.68)$$

The step error of Euler method is characterized by the remainder $O(\Delta t^2)$. However, with multiple application of the eq 6.68 the error will be accumulated and the order of the global error will be by a unit lower than the order of the local one, that is, Euler method is the method of first order.

There are many modifications of Euler method. One of them can be obtained when considering not two first summands in the expansion (eq 6.65) but three:

$$\bar{\mathbf{V}}(t+\Delta t)=\bar{\mathbf{V}}(t)+\Delta t\frac{d\bar{\mathbf{V}}(t)}{dt}+\frac{1}{2!}\Delta t^2\frac{d^2\bar{\mathbf{V}}(t)}{dt^2}+O(\Delta t^3) \quad (6.69)$$

The value of the first derivative is found by the correlation (eq 6.67). Having differentiated $\bar{\mathbf{V}}(t)$ by the formula of total derivative, we have the value of the second derivative:

$$\frac{d^2\bar{\mathbf{V}}(t)}{dt^2}=\frac{1}{m}\left[\frac{\partial\bar{\mathbf{F}}(t,\bar{\mathbf{r}}(t),\bar{\mathbf{V}}(t))}{\partial t}+\frac{\partial\bar{\mathbf{F}}(t,\bar{\mathbf{r}}(t),\bar{\mathbf{V}}(t))}{\partial\bar{\mathbf{r}}}\frac{d\bar{\mathbf{r}}(t)}{dt}\right]+$$

$$+\frac{1}{m}\frac{\partial\bar{\mathbf{F}}(t,\bar{\mathbf{r}}(t),\bar{\mathbf{V}}(t))}{\partial\bar{\mathbf{V}}}\frac{d\bar{\mathbf{V}}(t)}{dt} \quad (6.70)$$

The account of the first equation of the system (eq 6.64) and eq 6.67 produces the transformation of the second derivative as follows:

$$\frac{d^2\bar{\mathbf{V}}(t)}{dt^2}=\frac{1}{m}\left[\frac{\partial\bar{\mathbf{F}}(t,\bar{\mathbf{r}}(t),\bar{\mathbf{V}}(t))}{\partial t}+\frac{\partial\bar{\mathbf{F}}(t,\bar{\mathbf{r}}(t),\bar{\mathbf{V}}(t))}{\partial\bar{\mathbf{r}}}\bar{\mathbf{V}}(t)\right]+$$

$$+\frac{1}{m^2}\frac{\partial\bar{\mathbf{F}}(t,\bar{\mathbf{r}}(t),\bar{\mathbf{V}}(t))}{\partial\bar{\mathbf{V}}}\bar{\mathbf{F}}(t,\bar{\mathbf{r}}(t),\bar{\mathbf{V}}(t)) \quad (6.71)$$

The insertion of expressions for finding the first and second derivatives into the expansion (eq 6.69) allows obtaining the correlation to calculate the velocity on the step by time $t+\Delta t$:

$$\bar{\mathbf{V}}(t+\Delta t)=\bar{\mathbf{V}}(t)+\Delta t\frac{\bar{\mathbf{F}}(t,\bar{\mathbf{r}}(t),\bar{\mathbf{V}}(t))}{m}++\frac{1}{2!}\Delta t^2\frac{1}{m}\frac{\partial\bar{\mathbf{F}}(t,\bar{\mathbf{r}}(t),\bar{\mathbf{V}}(t))}{\partial t}+$$

$$+\frac{1}{2!}\frac{\Delta t^2}{m}\left[\frac{\partial\bar{\mathbf{F}}(t,\bar{\mathbf{r}}(t),\bar{\mathbf{V}}(t))}{\partial\bar{\mathbf{r}}}\bar{\mathbf{V}}(t)+\frac{\partial\bar{\mathbf{F}}(t,\bar{\mathbf{r}}(t),\bar{\mathbf{V}}(t))}{\partial\bar{\mathbf{V}}}\frac{\bar{\mathbf{F}}(t,\bar{\mathbf{r}}(t),\bar{\mathbf{V}}(t))}{m}\right]+O(\Delta t^3) \quad (6.72)$$

The method (eq 6.72) is called the modified Euler method. The step error of modified Euler method is determined by the value $O(\Delta t^3)$, consequently, the method has second order of accuracy. The disadvantage of this method is the necessity to calculate partial derivatives of the function $\bar{\mathbf{F}}(t,\bar{\mathbf{r}},\bar{\mathbf{V}})$.

If the value of function $\bar{\mathbf{F}}$ is used on the new time step $t+\Delta t$ in the eq 6.68, we have the expression for the implicit Euler method. The application of implicit Euler method is complicated by the fact that it is necessary to solve a nonlinear equation on each time step to calculate the value of $\bar{\mathbf{V}}(t+\Delta t)$. The method has first order of accuracy.

$$\bar{\mathbf{V}}(t+\Delta t)=\bar{\mathbf{V}}(t)+\Delta t\frac{\bar{\mathbf{F}}(t+\Delta t,\bar{\mathbf{r}}(t+\Delta t),\bar{\mathbf{V}}(t+\Delta t))}{m}+O(\Delta t^2) \quad (6.73)$$

The combination of explicit and implicit Euler method resulted in the emergence of trapezoidal method (eq 6.65). The trapezoidal method is referred to methods of second order of accuracy. It can be obtained with the help of quadrature formula of trapezoids.

$$\bar{\mathbf{V}}(t+\Delta t)=\bar{\mathbf{V}}(t)+\frac{\Delta t}{2}\frac{\bar{\mathbf{F}}(t,\bar{\mathbf{r}}(t),\bar{\mathbf{V}}(t))}{m}+$$

$$+\frac{\Delta t}{2}\frac{\bar{\mathbf{F}}(t+\Delta t,\bar{\mathbf{r}}(t+\Delta t),\bar{\mathbf{V}}(t+\Delta t))}{m}+O(\Delta t^3) \quad (6.74)$$

The modification of trapezoidal method by explicit Euler method resulted in the emergence of the hybrid method called Heun method:

$$\bar{\mathbf{V}}(t+\Delta t)=\bar{\mathbf{V}}(t)+\frac{\Delta t}{2}\frac{\bar{\mathbf{F}}(t,\bar{\mathbf{r}}(t),\bar{\mathbf{V}}(t))}{m}+$$

$$+\frac{\Delta t}{2}\frac{\overline{\mathbf{F}}\left(t+\Delta t,\overline{\mathbf{r}}(t+\Delta t),\overline{\mathbf{V}}(t)+\Delta t\overline{\mathbf{F}}\left(t,\overline{\mathbf{r}}(t),\overline{\mathbf{V}}(t)\right)\right)}{m}+O\left(\Delta t^3\right) \quad (6.75)$$

The higher accuracy can be achieved when applying several iterations by trapezoidal method on one time step. Such option of trapezoidal method is called the improved Euler–Cauchy method with iterative processing. The method principle is demonstrated by the formulas:

$$\overline{\mathbf{V}}^0\left(t+\Delta t\right)=\overline{\mathbf{V}}\left(t\right)+\Delta t\frac{\overline{\mathbf{F}}\left(t,\overline{\mathbf{r}}(t),\overline{\mathbf{V}}(t)\right)}{m}+O\left(\Delta t^2\right), \quad (6.76)$$

$$\overline{\mathbf{V}}^p\left(t+\Delta t\right)=\overline{\mathbf{V}}\left(t\right)+\frac{\Delta t}{2}\frac{\overline{\mathbf{F}}\left(t,\overline{\mathbf{r}}(t),\overline{\mathbf{V}}(t)\right)}{m}+$$

$$+\frac{\Delta t}{2}\frac{\overline{\mathbf{F}}\left(t+\Delta t,\overline{\mathbf{r}}(t+\Delta t),\overline{\mathbf{V}}^{p-1}(t+\Delta t)\right)}{m}+O\left(\Delta t^3\right) \quad (6.77)$$

The accuracy of Euler–Cauchy method with iterative processing is achieved by the coincidence of some number of orders of $\overline{\mathbf{V}}^{p-1}\left(t+\Delta t\right)$ and $\overline{\mathbf{V}}^{p}\left(t+\Delta t\right)$ on each time step, as well as the method accuracy order.

The integration of the eq 6.67 on the time interval $\left[t-\Delta t, t+\Delta t\right]$ and application of midpoint quadrature formula to calculate the integral of function $\overline{\mathbf{F}}$ allows obtaining the qualified Euler method:

$$\overline{\mathbf{V}}\left(t+\Delta t\right)=\overline{\mathbf{V}}\left(t-\Delta t\right)+2\Delta t\frac{\overline{\mathbf{F}}\left(t,\overline{\mathbf{r}}(t),\overline{\mathbf{V}}(t)\right)}{m}+O\left(\Delta t^3\right) \quad (6.78)$$

Qualified Euler method is the method of second order. The method feature is that for finding the solution on the step $t + \Delta t$, it is necessary to know the solution on t and $t - \Delta t$ time intervals, that is, the method comprises two steps.

6.3.2 *Runge–Kutta Methods*

The principle of Runge–Kutta methods of k order of accuracy is demonstrated in the use of the following formula:

$$\bar{V}(t+\Delta t)=\bar{V}(t)+\frac{\Delta t}{m}\bar{\mathbf{\phi}}\left(t,\bar{\mathbf{r}}(t),\bar{V}(t),\Delta t\right)+O\left(\Delta t^{k+1}\right), \qquad (6.79)$$

where $\bar{\varphi}\left(t,\bar{\mathbf{r}},\bar{V},\Delta t\right)$ = function approximating Taylor series of k order and not comprising partial derivatives of function $\bar{\mathbf{F}}$.

Multi-parameter $\bar{\varphi}\left(t,\bar{\mathbf{r}},\bar{V},\Delta t\right)$ is selected for Runge–Kutta methods. Its parameters are determined comparing with solution expansion into Taylor series.

In general, the family of Runge–Kutta methods of k order of accuracy is represented by the system (eq 6.71), where $i=2,3,\ldots,k$. The parameters of the system a_i,b_{ij},c_i are found from the solution expansion into Taylor series of the required order of accuracy:

$$\begin{cases} K_1(t)=\dfrac{\bar{\mathbf{F}}\left(t,\bar{\mathbf{r}}(t),\bar{\mathbf{V}}(t)\right)}{m}, \\ K_i(t)=\dfrac{1}{m}\bar{\mathbf{F}}\left(t+a_i\Delta t,\bar{\mathbf{r}}(t+a_i\Delta t),\bar{\mathbf{V}}(t)+\Delta t\sum_{j=1}^{i-1}b_{ij}K_j(t)\right), \\ \bar{\mathbf{V}}(t+\Delta t)=\bar{\mathbf{V}}(t)+\Delta t\sum_{i=1}^{k}c_iK_i(t)+O\left(\Delta t^{k+1}\right). \end{cases} \qquad (6.80)$$

Runge–Kutta methods of second order of accuracy and higher have the significant advantage in comparison with Euler methods. There is no need to calculate partial derivatives of function $\bar{\mathbf{F}}$ when developing these methods. Besides, Runge–Kutta methods are one-step ones.

Runge–Kutta method of fourth order is a special case and the most frequently used method in this family. The correlations describing this method are as follows:

$$\begin{cases} K_1(t)=\dfrac{\bar{\mathbf{F}}\left(t,\bar{\mathbf{r}}(t),\bar{\mathbf{V}}(t)\right)}{m}, \\ K_2(t)=\dfrac{1}{m}\bar{\mathbf{F}}\left(t+\dfrac{\Delta t}{2},\bar{\mathbf{r}}(t+\dfrac{\Delta t}{2}),\bar{\mathbf{V}}(t)+\dfrac{\Delta t}{2}K_1(t)\right), \end{cases} \qquad (6.81)$$

$$\begin{cases} K_3(t) = \frac{1}{m}\overline{\mathbf{F}}\left(t + \frac{\Delta t}{2}, \overline{\mathbf{r}}(t + \frac{\Delta t}{2}), \overline{\mathbf{V}}(t) + \frac{\Delta t}{2}K_2(t)\right), \\ K_4(t) = \frac{1}{m}\overline{\mathbf{F}}\left(t + \Delta t, \overline{\mathbf{r}}(t + \Delta t), \overline{\mathbf{V}}(t) + \Delta t K_3(t)\right), \\ \Delta\overline{\mathbf{V}}(t) = \frac{\Delta t}{6}\left(K_1(t) + 2K_2(t) + 2K_3(t) + K_4(t)\right), \\ \overline{\mathbf{V}}(t + \Delta t) = \overline{\mathbf{V}}(t) + \Delta\overline{\mathbf{V}}(t) + O\left(\Delta t^5\right). \end{cases} \tag{6.82}$$

6.3.3 Adams Methods

The integral-interpolational approach is applied to determine Adams numerical methods. Having integrated the eq 6.67 on the time interval $[t, t + \Delta t]$, the following expression is obtained:

$$\overline{\mathbf{V}}(t + \Delta t) = \overline{\mathbf{V}}(t) + \frac{1}{m}\int_{t}^{t+\Delta t} \overline{\mathbf{F}}\left(t, \overline{\mathbf{r}}(t), \overline{\mathbf{V}}(t)\right)dt \tag{6.83}$$

The sub-integral function $\overline{\mathbf{F}}(t, \overline{\mathbf{r}}(t), \overline{\mathbf{V}}(t))$ is substituted for the interpolating polynomial. Newton interpolation formula allows obtaining two types of polynomial for function $\overline{\mathbf{F}}$: when interpolating about point t and when interpolating about point $t + \Delta t$. Both variants of the interpolating polynomial comprise finite differences of function $\overline{\mathbf{F}}$ and serve as a basis for multi-step Adams methods.

When using the interpolating polynomial for function $\overline{\mathbf{F}}$ about point t, the family of Adams–Bashforth extrapolation methods is obtained. The order of accuracy changes depending on the number of steps of Adams–Bashforth methods. These methods refer to explicit methods of solving differential equations.

One-step Adams–Bashforth method of first order, also known as Euler method, is represented by the following formula:

$$\overline{\mathbf{V}}(t + \Delta t) = \overline{\mathbf{V}}(t) + \frac{\Delta t}{m}\overline{\mathbf{F}}\left(t, \overline{\mathbf{r}}(t), \overline{\mathbf{V}}(t)\right) + O\left(\Delta t^2\right), \tag{6.84}$$

two-step method of second order—

$$\bar{\mathbf{V}}(t+\Delta t)=\bar{\mathbf{V}}(t)+\frac{\Delta t}{2m}\left[3\bar{\mathbf{F}}\left(t,\bar{\mathbf{r}}(t),\bar{\mathbf{V}}(t)\right)-\right.$$

$$\left.-\bar{\mathbf{F}}\left(t-\Delta t,\bar{\mathbf{r}}(t-\Delta t),\bar{\mathbf{V}}(t-\Delta t)\right)\right]+O\left(\Delta t^{3}\right), \quad (6.85)$$

three-step method of third order—

$$\bar{\mathbf{V}}(t+\Delta t)=\bar{\mathbf{V}}(t)+\frac{\Delta t}{12m}\left[23\bar{\mathbf{F}}\left(t,\bar{\mathbf{r}}(t),\bar{\mathbf{V}}(t)\right)-16\bar{\mathbf{F}}\left(t-\Delta t,\bar{\mathbf{r}}(t-\Delta t),\bar{\mathbf{V}}(t-\Delta t)\right)+\right.$$

$$\left.+5\bar{\mathbf{F}}\left(t-2\Delta t,\bar{\mathbf{r}}(t-2\Delta t),\bar{\mathbf{V}}(t-2\Delta t)\right)\right]+O\left(\Delta t^{4}\right), \quad (6.86)$$

four-step method of fourth order—

$$\bar{\mathbf{V}}(t+\Delta t)=\bar{\mathbf{V}}(t)+\frac{\Delta t}{24m}\left[55\bar{\mathbf{F}}\left(t,\bar{\mathbf{r}}(t),\bar{\mathbf{V}}(t)\right)-\right.$$

$$-59\bar{\mathbf{F}}\left(t-\Delta t,\bar{\mathbf{r}}(t-\Delta t),\bar{\mathbf{V}}(t-\Delta t)\right)+37\bar{\mathbf{F}}\left(t-2\Delta t,\bar{\mathbf{r}}(t-2\Delta t),\bar{\mathbf{V}}(t-2\Delta t)\right)-$$

$$\left.-9\bar{\mathbf{F}}\left(t-3\Delta t,\bar{\mathbf{r}}(t-3\Delta t),\bar{\mathbf{V}}(t-3\Delta t)\right)\right]+O\left(\Delta t^{5}\right) \quad (6.87)$$

Adams–Moulton interpolation methods are based on the application of interpolating polynomial about point $t+\Delta t$ for function $\bar{\mathbf{F}}$ in the eq 6.10. Adams–Moulton methods are implicit methods.

One-step Adams–Moulton method of first order, also known as implicit Euler method, is written down as follows:

$$\bar{\mathbf{V}}(t+\Delta t)=\bar{\mathbf{V}}(t)+\frac{\Delta t}{m}\bar{\mathbf{F}}\left(t+\Delta t,\bar{\mathbf{r}}(t+\Delta t),\bar{\mathbf{V}}(t+\Delta t)\right)+O\left(\Delta t^{2}\right), \quad (6.88)$$

one-step method of second order, known as trapezoidal method—

$$\bar{\mathbf{V}}(t+\Delta t)=\bar{\mathbf{V}}(t)+\frac{\Delta t}{2m}\bar{\mathbf{F}}\left(t,\bar{\mathbf{r}}(t),\bar{\mathbf{V}}(t)\right)+$$

$$+\frac{\Delta t}{2m}\bar{\mathbf{F}}\left(t+\Delta t,\bar{\mathbf{r}}(t+\Delta t),\bar{\mathbf{V}}(t+\Delta t)\right)+O\left(\Delta t^{3}\right), \quad (6.89)$$

two-step method of third order—

$$\bar{\mathbf{V}}(t+\Delta t)=\bar{\mathbf{V}}(t)+\frac{\Delta t}{12m}\Big[5\bar{\mathbf{F}}\big(t+\Delta t,\bar{\mathbf{r}}(t+\Delta t),\bar{\mathbf{V}}(t+\Delta t)\big)+8\bar{\mathbf{F}}\big(t,\bar{\mathbf{r}}(t),\bar{\mathbf{V}}(t)\big)-$$

$$-\bar{\mathbf{F}}\big(t-\Delta t,\bar{\mathbf{r}}(t-\Delta t),\bar{\mathbf{V}}(t-\Delta t)\big)\Big]+O\left(\Delta t^{4}\right), \tag{6.90}$$

three-step method of fourth order—

$$\bar{\mathbf{V}}(t+\Delta t)=\bar{\mathbf{V}}(t)+\frac{\Delta t}{24m}\Big[9\bar{\mathbf{F}}\big(t+\Delta t,\bar{\mathbf{r}}(t+\Delta t),\bar{\mathbf{V}}(t+\Delta t)\big)+19\bar{\mathbf{F}}\big(t,\bar{\mathbf{r}}(t),\bar{\mathbf{V}}(t)\big)-$$

$$-5\bar{\mathbf{F}}\big(t-\Delta t,\bar{\mathbf{r}}(t-\Delta t),\bar{\mathbf{V}}(t-\Delta t)\big)+\bar{\mathbf{F}}\big(t-2\Delta t,\bar{\mathbf{r}}(t-2\Delta t),\bar{\mathbf{V}}(t-2\Delta t)\big)\Big]+O\left(\Delta t^{5}\right) \tag{6.91}$$

6.3.4 Predictor–Corrector Methods

Predictor–corrector methods are two-step methods combining the application of explicit and implicit methods. The assumed solution of $\bar{\mathbf{V}}^{P}(t+\Delta t)$, called "the predictor," is calculated on the first step. Then the predictor is corrected and the obtained value $\bar{\mathbf{V}}(t+\Delta t)$—corrector—is the desired solution.

The predictor and corrector algorithms based on Adams methods are widely spread. The predictor is calculated by explicit Adams–Bashforth methods, and the corrector—using Adams–Moulton formulas. Due to the fact that Adams–Bashforth and Adams–Moulton methods can have different orders of accuracy, the following Adams predictor–corrector methods are possible:

of first order (explicit–implicit Euler method)—

$$\begin{cases}\bar{\mathbf{V}}^{P}(t+\Delta t)=\bar{\mathbf{V}}(t)+\dfrac{\Delta t}{m}\bar{\mathbf{F}}\big(t,\bar{\mathbf{r}}(t),\bar{\mathbf{V}}(t)\big)+O\left(\Delta t^{2}\right),\\ \bar{\mathbf{V}}(t+\Delta t)=\bar{\mathbf{V}}(t)+\dfrac{\Delta t}{m}\bar{\mathbf{F}}\big(t+\Delta t,\bar{\mathbf{r}}(t+\Delta t),\bar{\mathbf{V}}^{P}(t+\Delta t)\big)+O\left(\Delta t^{2}\right),\end{cases} \tag{6.92}$$

of second order—

$$\begin{cases} \bar{\mathbf{V}}^P(t+\Delta t) = \bar{\mathbf{V}}(t) + \dfrac{\Delta t}{2m}\Big[3\bar{\mathbf{F}}\big(t,\bar{\mathbf{r}}(t),\bar{\mathbf{V}}(t)\big) - \\ \qquad -\bar{\mathbf{F}}\big(t-\Delta t,\bar{\mathbf{r}}(t-\Delta t),\bar{\mathbf{V}}(t-\Delta t)\big)\Big] + O\big(\Delta t^3\big), \\ \bar{\mathbf{V}}(t+\Delta t) = \bar{\mathbf{V}}(t) + \dfrac{\Delta t}{2m}\Big[\bar{\mathbf{F}}\big(t+\Delta t,\bar{\mathbf{r}}(t+\Delta t),\bar{\mathbf{V}}^P(t+\Delta t)\big) + \\ \qquad +\bar{\mathbf{F}}\big(t,\bar{\mathbf{r}}(t),\bar{\mathbf{V}}(t)\big)\Big] + O\big(\Delta t^3\big), \end{cases} \tag{6.93}$$

of third order—

$$\begin{cases} \bar{\mathbf{V}}^P(t+\Delta t) = \bar{\mathbf{V}}(t) + \dfrac{\Delta t}{12m}\Big[23\bar{\mathbf{F}}\big(t,\bar{\mathbf{r}}(t),\bar{\mathbf{V}}(t)\big) - 16\bar{\mathbf{F}}\big(t-\Delta t,\bar{\mathbf{r}}(t-\Delta t),\bar{\mathbf{V}}(t-\Delta t)\big) + \\ \qquad +5\bar{\mathbf{F}}\big(t-2\Delta t,\bar{\mathbf{r}}(t-2\Delta t),\bar{\mathbf{V}}(t-2\Delta t)\big)\Big] + O\big(\Delta t^4\big), \\ \bar{\mathbf{V}}(t+\Delta t) = \bar{\mathbf{V}}(t) + \dfrac{\Delta t}{12m}\Big[5\bar{\mathbf{F}}\big(t+\Delta t,\bar{\mathbf{r}}(t+\Delta t),\bar{\mathbf{V}}^P(t+\Delta t)\big) + 8\bar{\mathbf{F}}\big(t,\bar{\mathbf{r}}(t),\bar{\mathbf{V}}(t)\big) - \\ \qquad -\bar{\mathbf{F}}\big(t-\Delta t,\bar{\mathbf{r}}(t-\Delta t),\bar{\mathbf{V}}(t-\Delta t)\big)\Big] + O\big(\Delta t^4\big). \end{cases} \tag{6.94}$$

The predictor–corrector methods of fourth order of accuracy can be developed by analogy with (eqs 6.92–6.94), guided by Adams–Bashforth (eq 6.87) and Adams–Moulton eq 6.91.

Comparing the solutions of $\bar{\mathbf{V}}^P(t+\Delta t)$ and $\bar{\mathbf{V}}(t+\Delta t)$ in predictor–corrector methods, it is possible to control the error value that is a considerable advantage of these methods.

6.3.5 Verlet algorithm

Numerical Verlet algorithms, which are the adaptations of explicit and implicit Euler methods, are most widely spread among the numerical algorithms of molecular dynamics. Several variations of Verlet method are known, differing in accuracy order and development of numerical algorithms.

Let us consider the expansion of the particle spatial vector into Taylor series at the time moments $t + \Delta t$ and $t - \Delta t$. These expansions for i atom are represented by the corresponding formulas[20,21]:

$$\bar{\mathbf{r}}_i(t+\Delta t) = \bar{\mathbf{r}}_i(t) + \bar{\mathbf{V}}_i(t)\Delta t + \frac{\bar{\mathbf{F}}_i(t,\bar{\mathbf{r}}(t))}{2m_i}\Delta t^2 + \frac{\Delta t^3}{3!}\frac{d}{dt}\left(\frac{\bar{\mathbf{F}}_i(t,\bar{\mathbf{r}}(t))}{m_i}\right) + O\big(\Delta t^4\big), \tag{6.95}$$

$$\bar{\mathbf{r}}_i(t-\Delta t) = \bar{\mathbf{r}}_i(t) - \bar{\mathbf{V}}_i(t)\Delta t + \frac{\bar{\mathbf{F}}_i(t,\bar{\mathbf{r}}(t))}{2m_i}\Delta t^2 - \frac{\Delta t^3}{3!}\frac{d}{dt}\left(\frac{\bar{\mathbf{F}}_i(t,\bar{\mathbf{r}}(t))}{m_i}\right) + O\left(\Delta t^4\right) \quad (6.96)$$

The position of i atom on the new time layer is calculated summing up the eqs 6.95 and 6.96:

$$\bar{\mathbf{r}}_i(t+\Delta t) = 2\bar{\mathbf{r}}_i(t) - \bar{\mathbf{r}}_i(t-\Delta t) + \frac{\bar{\mathbf{F}}_i(t,\bar{\mathbf{r}}(t))}{m_i}\Delta t^2 + O\left(\Delta t^4\right) \quad (6.97)$$

Thus, to find the spatial coordinates on the new time layer, it is necessary to know the positions on the current and previous steps by time and value of force $\bar{\mathbf{F}}_i(t,\bar{\mathbf{r}}(t))$. It is not necessary to calculate the values of particle velocities, but it is possible to use the difference scheme of the central derivative, if required:

$$\bar{\mathbf{V}}_i(t) = \frac{\bar{\mathbf{r}}_i(t+\Delta t) - \bar{\mathbf{r}}_i(t-\Delta t)}{2\Delta t} + O\left(\Delta t^2\right) \quad (6.98)$$

In the right side of the eq 6.97, the first two summands have the order of small quantity $O\left(\Delta t^0\right)$, and the third—$O\left(\Delta t^2\right)$. The difference in the orders of small quantity results in the accumulation of round-off error and decrease in the accuracy scheme. Therefore, the modifications of the basic Verlet algorithm are spread: velocity Verlet algorithm, Leapfrog algorithm.

Leapfrog algorithm is a method of second order of accuracy, and it consumes less computational resources in comparison with the basic version of Verlet algorithm. With this algorithm the velocity of particles is calculated on intermediary $\left(t+\frac{1}{2}\Delta t\right)$ steps by time. The following correlation is used to calculate the velocities on the next step:

$$\bar{\mathbf{V}}_i\left(t+\frac{1}{2}\Delta t\right) = \bar{\mathbf{V}}_i\left(t-\frac{1}{2}\Delta t\right) + \Delta t\frac{\bar{\mathbf{F}}_i(t,\bar{\mathbf{r}}(t))}{m_i} \quad (6.99)$$

The positions of particles are found as follows:

$$\bar{\mathbf{r}}_i(t+\Delta t) = \bar{\mathbf{r}}_i(t) + \bar{\mathbf{V}}_i\left(t+\frac{1}{2}\Delta t\right)\Delta t \quad (6.100)$$

The velocity values for each particle at the current time moment can be obtained from the following expression:

$$\bar{\mathbf{V}}_i(t) = \frac{1}{2}\left[\bar{\mathbf{V}}_i\left(t + \frac{1}{2}\Delta t\right) + \bar{\mathbf{V}}_i\left(t - \frac{1}{2}\Delta t\right)\right] \tag{6.101}$$

Velocity Verlet algorithm also possesses second order of accuracy.[20] This method is rather economical in relation to computational resources and is one-step, that is, there is no need to calculate the initial time layers with the help of other methods when using it. The following sequence of operations is followed when applying velocity Verlet algorithm:

(1) Setting up initial conditions (eq 6.11).

(2) Calculation of new positions of atoms for the time moment $t + \Delta t$:

$$\bar{\mathbf{r}}_i(t + \Delta t) = \bar{\mathbf{r}}_i(t) + \bar{\mathbf{V}}_i(t)\Delta t + \frac{\bar{\mathbf{F}}_i(t, \bar{\mathbf{r}}(t))}{2m_i}\Delta t^2 \tag{6.102}$$

(3) Calculation of velocities on the half-step by time.

$$\bar{\mathbf{V}}_i\left(t + \frac{1}{2}\Delta t\right) = \bar{\mathbf{V}}_i(t) + \Delta t\frac{\bar{\mathbf{F}}_i(t, \bar{\mathbf{r}}(t))}{2m_i} \tag{6.103}$$

(4) Recalculation of forces acting upon the atoms $\bar{\mathbf{F}}_i\left(t + \Delta t, \bar{\mathbf{r}}(t + \Delta t)\right)$ through the above $\bar{\mathbf{r}}_i(t + \Delta t)$ by the eq 6.12, taking into account the potential used.

(5) Finding the velocities at the time moment $t + \Delta t$:

$$\bar{\mathbf{V}}_i\left(t + \Delta t\right) = \bar{\mathbf{V}}_i(t + \frac{1}{2}\Delta t) + \Delta t\frac{\bar{\mathbf{F}}_i(t + \Delta t, \bar{\mathbf{r}}(t + \Delta t))}{2m_i} \tag{6.104}$$

(6) Calculation of thermodynamic parameters, analysis of the system state, if necessary.

(7) Transition to the next step by time.

The velocities and position of particles of the nanosystem are calculated at any time moment as a result of implementation of this algorithm. The shortcomings of basic Verlet algorithm are eliminated in velocity Verlet algorithm. Velocity Verlet algorithm was applied when solving the problem of modeling the processes of obtaining special nanofilm coatings on the templates of porous aluminum oxide.

6.4 THERMODYNAMIC PARAMETERS AND ENERGY OF NANOSYSTEM

6.4.1 General Thermodynamic and Energy Parameters

The potential energy of the molecule and system of the formation of special nanofilm coatings on the templates of porous aluminum oxide is, in general, of electromagnetic nature and is given in the form of potentials. It is frequently necessary to calculate the kinetic energy of the system. The expression for the average kinetic energy of the system at time moment t is as follows[10]:

$$E(t)=\frac{\sum_{i=1}^{N} m_i\left(\bar{\mathbf{V}}_i(t)\right)^2}{2N}, \tag{6.105}$$

where m_i = molecular mass of i atom;

$\bar{\mathbf{V}}_i(t)$ = atom velocity;

N = total number of atoms.

The temperature of molecular system is closely connected with kinetic energy. The following ratio is know from the course of physics:

$$E=\frac{3k_{\mathrm{B}}T}{2}, \tag{6.106}$$

where k_B = Boltzmann constant.

The eqs 6.105 and 6.106 allow calculating the instantaneous value of temperature. Thus, in nanosystems the instantaneous temperature of the molecular system is defined as the average kinetic energy of the system[55,56]:

$$T(t)=\frac{1}{3Nk_{\mathrm{B}}}\sum_{i=1}^{N} m_i\left(\bar{\mathbf{V}}_i(t)\right)^2 \tag{6.107}$$

When investigating nanostructured photoelectric converters, the influence of the initial temperature on the distribution of molecule velocities is the important moment influencing the substantiality of the processes flowing in the nanosystem. In the problems on the formation of nanosized elements, the velocity field at the initial time moment was selected in accordance with Maxwell distribution. Maxwell distribution for velocity vector $\bar{\mathbf{V}} = (V_x, V_y, V_z)$ is the product of distributions for each of three directions:

$$f_V(V_x, V_y, V_z) = f_V(V_x) f_V(V_y) f_V(V_z), \tag{6.108}$$

where the distribution along one direction is found by the following correlation:

$$f_V(V_j) = \sqrt{\frac{m_i}{2\pi k_B T}} \exp\left[\frac{-m_i V_j^2}{2k_B T}\right], \; j = \{x, y, z\} \tag{6.109}$$

This distribution has the form of the normal distribution. As it should be expected, the average velocity for the gas at rest in any direction equals zero.

The following expression is used to calculate the nanosystem pressure:

$$P(t) = \frac{1}{3W}\left[\sum_{i=1}^{N} m_i\left(\bar{\mathbf{V}}_i(t)\right)^2 - \sum_{\substack{i,j \\ i<i}} (\bar{\mathbf{r}}_{\tilde{j}}(t) - \bar{\mathbf{r}}_i(t))\bar{\mathbf{F}}_{i\tilde{j}}(t)\right], \tag{6.110}$$

where W is volume occupied by the nanosystem.

The second summand in eq 6.110 is responsible for the pair interaction of atoms. All images of j atom are considered and the interaction of i atom with the nearest of $\tilde{j}$ is calculated together with the pair of i and j atoms. Function $\bar{\mathbf{F}}_{i\tilde{j}}(t)$ characterizes the interaction value between the atoms.

Special algorithms—thermostats—are used to take into account the energy exchange effects with the environment.[57–60] The use of thermostat allows calculating the parameters of the system modeled by the methods of molecular dynamics at constant temperature of the environment, or, on the contrary, change the environmental temperature following a certain law.

There are several types of thermostats. For example: collision thermostat, Berendsen thermostat, friction thermostat, and Nose–Hoover thermostat.

The use of thermostat is especially important at the stage of system relaxation. In case of thermodynamic equilibrium, the thermostat temperature and average temperature of the molecular system need to coincide. The energies of subsystems are usually much less than the thermostat energy, this is the condition of practical equilibrium. When studying molecular dynamics, the thermostat temperature is usually registered. At the same time, the temperature of molecular system can change due to different reasons. For example, due to the finite integration step the particle can be found in classically forbidden region. This results in the energy jump followed by the temperature jump.

6.4.2 Collision Thermostat

In this thermostat model, the medium of virtual particles, interacting with the particles of the molecular system studied, is introduced.[61,62] The collisions occur following the law of elastic balls. Varying the mass of virtual particles and collision frequency with the system atoms, it is possible to achieve the best agreement with the experimental data. When calculating under vacuum, the mass of virtual particles is usually set as 18 a.m.u., and collision frequency—55–60 ps^{-1}. By its viscous properties such medium is close to water under normal conditions.

The thermostat temperature defines the function of virtual particle distribution by velocities:

$$f(V) = \left(\frac{m}{2\pi k_B T}\right)^{\frac{3}{2}} V^2 \exp\left(-\frac{mV^2}{2k_B T}\right), \tag{6.111}$$

where $f(V)$ = function of probability distribution of virtual particles by velocities;

$f(V)dV$ = probability that the absolute velocity value of the virtual particle is in the interval between V and $V + dV$;

m = mass of virtual particle;

k_B = Boltzmann constant;

T = thermostat temperature.

The collision frequency is given by Poisson distribution:

$$p(r) = \left(\frac{\xi t}{r!} \right)^r e^{-\xi t}, \quad (6.112)$$

where $p(r)$ = probability that r collisions will take place during the time interval $(0,t)$;

ξ = collision frequency.

6.4.3 Berendsen Thermostat

Berendsen algorithm is based on the introduction of alternating-sign friction.[58] The temperature behavior when using Berendsen thermostat is demonstrated in Figure 6.11. The temperature deviations (T) from its equilibrium value (T_0) are corrected in compliance with Landau–Teller equation[63]:

$$\frac{dT(t)}{dt} = \frac{T_0 - T(t)}{\tau}, \quad (6.113)$$

where $T(t)$ = current value of the temperature.

The deviations in the temperature values decrease exponentially with the characteristic time τ. The change in the kinetic energy is modeled, rescaling seating the atom velocities in the molecular system on each step:

$$\lambda = \sqrt{1 + \frac{\Delta t}{\tau_1} \left(\frac{T_0}{T\left(t - \frac{\Delta t}{2}\right)} - 1 \right)}, \quad (6.114)$$

where λ = coefficient of velocity recalculation;

τ_1 = time constant (about 1 ps).

It is known that the application of Berendsen thermostat, especially for relatively small systems and at long trajectories, results in physically incorrect results due to the uneven distribution of energy by degrees of freedom.[64,65]

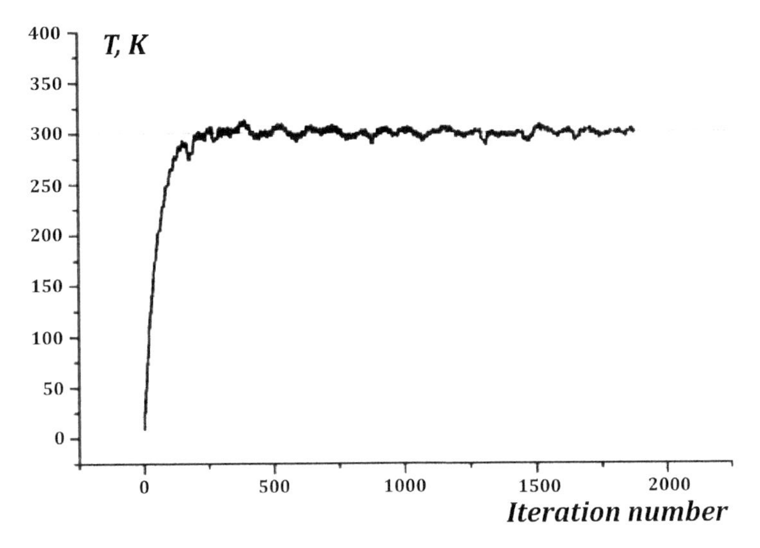

FIGURE 6.11 Characteristic behavior of the system temperature with Berendsen thermostat.

6.4.4 *Nose–Hoover Thermostat*

Berendsen thermostat provides the exponential dying-out of temperature to the required value. However, the following temperature fluctuations are great, therefore, Nose–Hoover thermostat is frequently used for further thermalization.[66,67]

Thermalization with the help of this method consists in the introduction of effective friction forces, proportional to the particle velocities with dynamically changing coefficient ξ, into the system.

$$\frac{d^2 r_i}{dt^2} = \frac{F_i}{m_i} - \xi \frac{dr_i}{dt} \tag{6.115}$$

The equations for coefficient ξ are solved by numerical integration by time together with the integration of motion equations:

$$\frac{d\xi}{dt} = \frac{1}{Q}\left(T - T_0\right) \tag{6.116}$$

Euler scheme of first order was used to solve the latter differential equation regarding ξ. The temperature behavior when using Nose–Hoover thermostat is given in Figure 6.12.

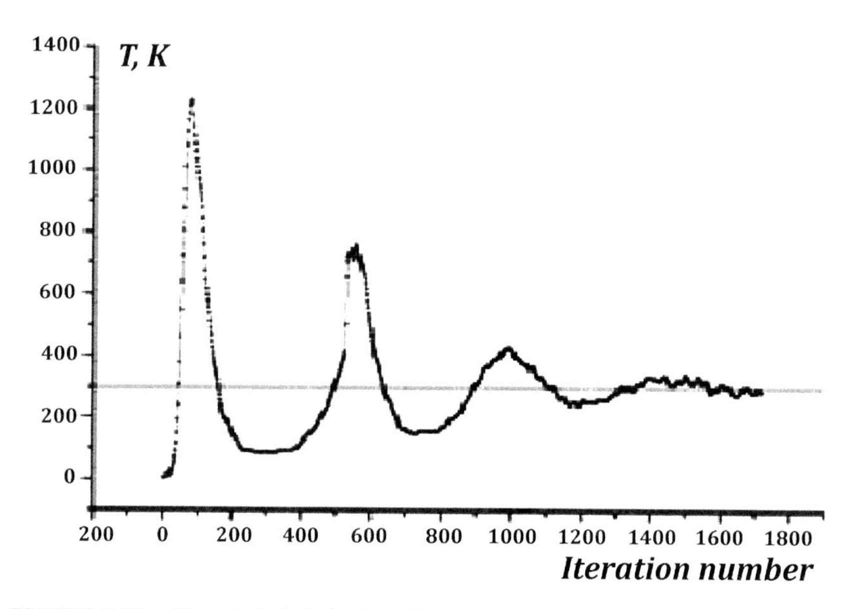

FIGURE 6.12 Characteristic behavior of the system temperature with Nose–Hoover thermostat.

Mass coefficient Q determines the rate of reaching the required temperature. This parameter can be found either by intuition or by devising a setup via other values. The convergence to temperature T_0 is of the form of oscillations with period τ_T.

$$Q = \frac{\tau_T^2 T_0}{4\pi^2} \tag{6.117}$$

6.4.5 *Friction Thermostat*

Here, in contrast to the previous thermostat, the friction force coefficient G_i is not found by integration, that is, it does not depend on the prehistory:

$$G_i = b\left(\frac{T_0}{T} - 1\right)\frac{d\,r_i}{dt}. \qquad (6.118)$$

Another peculiarity is that, in contrast to the previous cases, this thermalization is inversely proportional to the current temperature T. The force constant b is found experimentally. The characteristic behavior of the thermostat is given in Figure 6.13.

The use of barostat algorithms gives the possibility to model the system behavior under constant pressure. The main barostat principle is the change in the integration cell sizes until the required pressure is reached.[68] The methods of barostat operation used in the program are the same as in case of thermostats, only not velocities but coordinates are changed.

The pressure in molecular dynamics is calculated as follows:

$$W_1 = \sum_{i<j}^{N} F_{ij}\, r_{ij} \qquad (6.119)$$

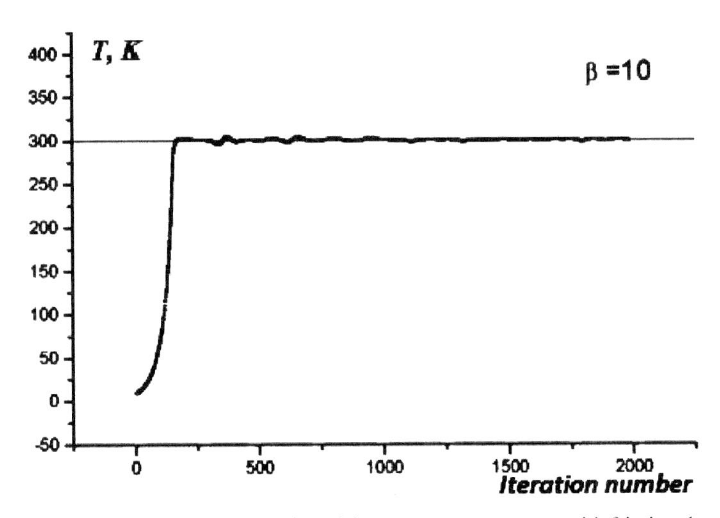

FIGURE 6.13 Characteristic behavior of the system temperature with friction thermostat.

After the calculation (eq 6.119), the pressure is found as:

$$P = \frac{2 \cdot K + W_1}{3 \cdot W}, \qquad (6.120)$$

where K = full kinetic energy;
W = volume.

Physically W_1 defines the "nonideality" degree of the molecule system, being responsible for their interaction.

6.4.6 Berendsen Barostat

Berendsen barostat allows scaling coordinates of atoms and vectors of atom velocities at each step by time with the help of matrix μ, which has the effect of first-order kinetic relaxation of pressure relative to pressure P_0:

$$\frac{dP}{dt} = \frac{P_0 - P}{\tau_p} \tag{6.121}$$

Matrix μ being scaled is as follows:

$$\mu = \delta_{ij} - \frac{\Delta t}{3\tau_p}\beta_{ij}\left\{P_{0ij} - P_{ij}(t)\right\}, \tag{6.122}$$

where β = defines isothermal compressibility of the system.

In most cases, β is the diagonal matrix with equal elements on the diagonal, whose values are usually unknown. Sometimes, it is sufficient to take the rough estimate, as the values of β influence only non-critical time constants of pressure during relaxation.

6.4.7 Parrinello–Rahman Barostat

In many respects, the operation algorithm of this barostat is similar to Nose–Hoover thermalization method but with respect to coordinates. The sizes of integration cell in this algorithm will change dynamically.[69]

The diagonal matrix b with geometric sizes of the integration cell $\{L_x, L_y, L_z\}$ located on the main diagonal is considered. The components of this matrix, that is, the cell sizes, will change dynamically and the particles will tend to get together or, on the contrary, will fly apart under the action of additional force depending on b. Matrix b is subjected to the following equation:

$$\frac{d^2b}{dt^2} = WW_2^{-1}b^{-1}\left(P - P_0\right) \tag{6.123}$$

Parrinello–Rahman and Nose–Hoover barostats can be combined with each other. They are connected as follows:

$$\frac{d^2r_i}{dt^2} = \frac{F_i}{m_i} - M\frac{dr_i}{dt}, \tag{6.124}$$

$$M = b^{-1}\left[b\frac{db'}{dt} + \frac{db}{dt}b'\right]b'^{-1}, \tag{6.125}$$

where W_2 = parameter, which determines the convergence rate to the given pressure P_0;

W = volume; M = dynamic coefficient.

Parameter W_2 is expressed via the medium compressibility β, relaxation time τ_p, and maximum element L of matrix b:

$$W = \frac{4\pi^2\beta_{ij}}{3\tau_p^2 L} \tag{6.126}$$

The values $\frac{db}{dt}$ and b were calculated with the successive application of Euler scheme of first order.

6.5 PERIODIC BOUNDARY CONDITIONS

The important feature of modeling nanofilm coatings on the templates of porous aluminum oxide is that the size of computational region is always finite. This results in the necessity to adequately select the boundary conditions when modeling the nanosystem behavior.

In general, the selection of one or another boundary condition is specified by physical setting up of the problem. Periodic boundary conditions are most common when modeling the infinite uniform medium. In this connection, the computational cell shape, in the simplest implementation, is taken as rectangular parallelepiped. When during the calculation the particle gets outside the computational region, its coordinates are automatically shifted in the required direction, as a result, it gets into the cell from another side. In this case, the particle velocity does not change. The forces acting upon each particle near the cell boundary are also calculated taking into account the availability of particle images from its other end.

Under the particle image we understand the virtual particle with coordinates obtained from the coordinates of the initial particle at its shifting by vector $S = \{S_x, S_y, S_z\}$, where $S_i = 0 \pm L_i$, L_i—size of the cell along the selected direction. Provided that the computational region is of rectangular shape, each particle has 26 images. When calculating the interaction between two particles, the so-called nearest image rule is applied, namely, the pair is produced either between the particles themselves or between the particle and nearest image of its interaction partner, depending on the ratio between the distances. This condition imposes an additional limitation on the correlation between the computational region size and truncation radius of the paired interaction: if L_x, L_y, L_z—cell sizes in three dimensions, then the following inequalities need to be fulfilled: $L_x > \frac{R_c}{2}, L_y > \frac{R_c}{2}, L_z > \frac{R_c}{2}$.

The reverse situation can arise together with the infinite medium modeling, when it becomes necessary to consider the closed finite system of interacting particles. In this case, an additional harmonic potential returning the ejected particles back to the computational cell is introduced outside the computational region. Here, the analogy with "rubber walls" is appropriate.

Let us designate the value of the particle output outside the computational region in each of directions as $\Delta X, \Delta Y, \Delta Z$. Then the interaction with the cell boundary is described with the help of potential:

$$E = \frac{1}{2} E_b \left(\left(\frac{\Delta X}{l_x} \right)^2 + \left(\frac{\Delta Y}{l_y} \right)^2 + \left(\frac{\Delta Z}{l_z} \right)^2 \right), \tag{6.127}$$

where $l_x = \frac{L_x}{N_x}, l_y = \frac{L_y}{N_y}, l_z = \frac{L_z}{N_z}$ is cell partition step into sub-regions to search for neighbors in three dimensions.

Parameter E_b with the energy dimensionality defines the rigidity of boundary conditions. The main requirement to the rigidity consists in the fact that the output of particles outside the computational region was small in comparison with the characteristic mesh step dividing the computational cell into sub-regions to search for neighbors. At the same time, the rigidity of boundary conditions should be limited from the top, as otherwise the integration step by time may be not sufficient for correct account of the interaction with the boundary.

Other situations also exist apart from the mentioned settings of the problem. A small heterogeneous system placed in the infinite medium (e.g., protein molecule in water) can be taken as an example. From physical point of view, the computational cell of spherical shape could be the most appropriate variant to describe the system under these conditions (just so the problem isotropy can be accounted at a large distance from the heterogeneous region). There are several ways to set up boundary conditions in this situation. First, the way similar, to some extent, to the implementation of periodic boundary conditions can be chosen, namely, to launch the excited particle back inside the computational cell but from another direction. Second, it is possible to eliminate the particles that excited the computational region from the consideration and to generate instead the flow of new particles at the boundary directed from the boundary inside the cell with set up parameters of distribution by velocities and directions. Third, it is possible to prohibit the particles to leave the computational region by specially setting up the reaction force from the cell boundary. The first two ways of implementation of spherically symmetrical boundary conditions operate well in gases.

In this work, Born–Karman periodic boundary conditions were selected as the boundary conditions, the scheme is given in Figure 6.14. In accordance with Born–Karman conditions, it is required for the function to be periodical along each of coordinates. In 3D case:

$$f(x+L,y,z)=f(x,y,z), f(x,y+L,z)=f(x,y,z), f(x,y,z+L)=f(x,y,z) \quad (6.128)$$

Periodic boundary conditions allow decreasing the size of computational volume and significantly reducing the computational costs during modeling.

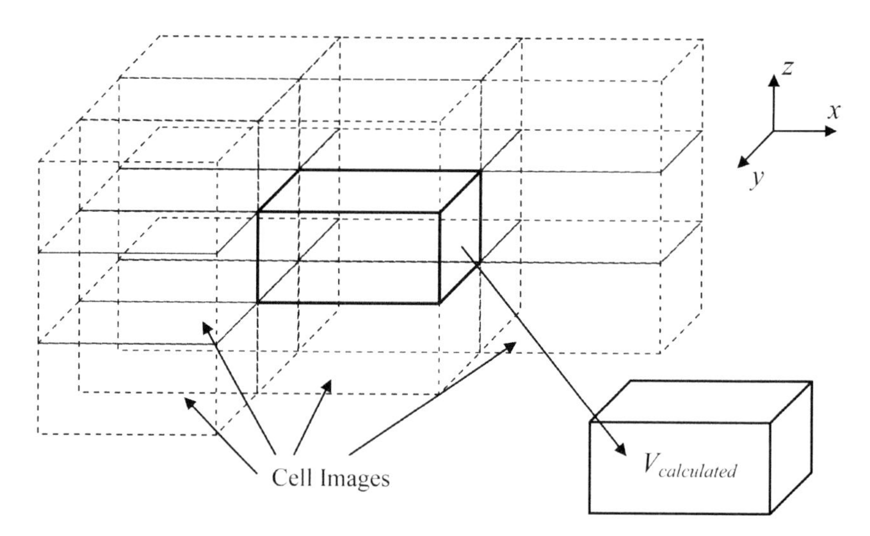

FIGURE 6.14 Use of periodic boundary conditions.

6.6 CONCLUSION

The chapter presents the fundamentals of computer modeling of the processes of obtaining nanostructured layers on the templates of porous aluminum oxide. The methods of modeling nanofilmed coatings contain the main equations of nanoobject motion, algorithms of temperature and pressure maintaining in the system, mechanisms of forming initial velocities, and coordinates connected with thermodynamic parameters of nanostructured objects. Periodic boundary conditions of the problem of forming specialized nanosized structures, which allow spreading the results of modeling the computational region in space over the whole investigated object are described.

Different types of potentials and force fields, which can be used when solving the problems of modeling the formation of nanofilms on the templates of porous aluminum oxide are given. The advantages and disadvantages of different types of potentials were analysed. It is shown that many-body potentials more accurately reproduce the properties of complex nanostructures.

The chapter contains the algorithms of numerical analysis of equations of motion and interaction of nanoparticles. Numerical methods allow modeling the behavior of optic and semiconductor nanosystems with different approximation orders and accuracy degrees.

KEYWORDS

- **mathematical modeling**
- **quantum mechanics**
- **embedded atom method**
- **Abell-Tersoff potential**
- **Stillinger-Weber potential**
- **modified embedded atom method**
- **thermostats and baristas in nanosystems**

REFERENCES

1. Dirac, P. A. M. *Principles of Quantum Mechanics;* Nauka: Moscow, 1979; p 408.
2. Landau, L. D.; Lifshitz, E. M. *Quantum Mechanics;* Nauka: Moscow, 1972; p 368.
3. Feynman, R.; Hibs, A. *Quantum Mechanics and Integrals Along Trajectories;* Mir: Moscow, 1968; p 382.
4. Fock, V. A. *The Beginnings of Quantum Mechanics;* Nauka: Moscow, 1976; p 376.
5. Gubanov, V. A.; Zhukov, V. P.; Litinsky, A. O. *Semiempirical Methods of Molecular Orbitals in Quantum Chemistry;* Nauka: Moscow, 1976; p 219.
6. Minkin, V. I. *Theory of Molecular Structure;* Phoenix: Rostov-on-Don, 1997; p 560.
7. Sigal, J. *Semi-Empirical Methods for Calculating the Electronic Structure;* Mir: Moscow, 1980; p 704.
8. Marx, D.; Hutter, J. Ab Initio Molecular Dynamics: Theory and Implementation. *Modern Methods and Algorithms of Quantum Chemistry,* Grotendorst, J., Ed.; NIC Series, Forschungszentrum Jülich, Germany, 2000, Vol. 1, pp 301–449.
9. Berlin, A. A.; Balabaev, N. K. Simulation of Properties of Solids and Liquids by Computer Simulation Methods. *Soros Educ. J.* **1997,** *11*, 85–92.
10. Romanova, T. A.; Krasnov, P. O.; Kachin, S. V.; Avramov, P. V. *Theory and Practice of Computer Modeling of Nanoobjects: Reference Book;* IPC of the KSTU: Krasnoyarsk, 2002; p 223.
11. Cagın, T.; Che, J.; Qi, Y.; Zhou, Y.; Demiralp, E.; Gao, G.; Goddard, W. A. Computational Materials Chemistry at the Nanoscale. *J. Nanoparticle Res.* 1999, *1*, 51–69.
12. Cornell, W. D.; Cieplak, P.; Bayly, C. I.; et al. A Second Generation Force Field for the Simulation of Proteins, Nucleic Acids and Organic Molecules. *J. Am. Chem. Soc.* 1995, *117* (19), 5179–5197.

13. Damm, W.; Frontera, A.; Tirado-Rives, J.; et al. OPLS All-Atom Force Field for Carbohydrates. *J. Comp. Chem.* 1997, *18*, 1955–1970.
14. Jorgensen, W. L.; Maxwell, D. S.; Tirado-Rives, J. Development and Testing of the OPLS All-Atom Force Field on Conformational Energetics and Properties of Organic Liquids. *J. Am. Chem. Soc.* 1996, *118* (45), 11225–11236.
15. Weiner, P. K.; Kollman, P. A. AMBER: Assisted Model Building with Energy Refinement. A General Program for Modeling Molecules and Their Interactions. *J. Comp. Chem.* 1981, *2* (3), 287–303.
16. Weiner. S. J.; Kollman, P. A.; Nguyen, D. T.; et al. An All Atom Force Field for Simulations of Proteins and Nucleic Acids. *J. Comp. Chem.* 1986, *7* (2), 230–252.
17. Halgren, T. A. Merk Molecular Force Field. 1. Basis, Form, Scope, Parametrization and Performance of MMFF94. *J. Comp. Chem.* 1996, *17* (5–6), 490–519.
18. Lin, V. S. -Y.; Nieweg, J. A.; Kern, C.; Trewyn, B. G.; Wiench, J. W.; Pruski, M. Acid-Base Mesoporous Calcia-Silica Catalysts for Cooperative Conversion of Bio-Based Feedstocks Into Biodiesel. *Prepr. Symp. Am. Chem. Soc. Div. Fuel Chem.* 2006, *51*, 426–427.
19. Schuler, L. D.; Daura, X.; Gunsteren, W. F. An Improved GROMOS96 Force Field for Aliphatic Hydrocarbons in the Condensed Phase. *J. Comp. Chem.* 2001, *22* (11), 1205–1218.
20. Verlet, L. Computer "Experiments" on Classical Fluids. I. Thermodynamical Properties of Lennard-Jones Molecules. *Phys. Rev.* 1967, *159* (1), 98–103.
21. Verlet, L. Computer "Experiments" on Classical Fluids. II. Equilibrium Correlation Functions. *Phys. Rev.* 1968, *165* (1), 201–214.
22. Daw, M. S.; Baskes, M. I. Embedded-Atom Method: Derivation and Application to Impurities, Surfaces, and Other Defects in Metals. *Phys. Rev. B* 1984, *29* (12), 6443–6453.
23. Ercolessi, F. A. *Molecular Dynamics Primer;* Spring College in Computational Physics, ICTP: Trieste, 1997; p 48.
24. Stillinger, F. H.; Weber, T. A. Computer Simulation of Local Order in Condensed Phases of Silicon. *Phys. Rev. B* 1985, *31* (8), 5262–5271.
25. Tersoff, J. New Empirical Approach for the Structure and Energy of Covalent Systems. *Phys Rev B* 1988, *37* (12), 6991–7000.
26. Venezuela, P.; Tersoff, J. Alloy Decomposition During Growth due to Mobility Differences. *Phys Rev B* 1998, *58* (16), 10871–10874.
27. Daw, M. S.; Baskes, M. I. Semiempirical, Quantum Mechanical Calculations of Hydrogen Embrittlement in Metals. *Phys. Rev. Lett.* 1983, *50* (17), 1285–1288.
28. Daw, M. S. Model of Metallic Cohesion: The Embedded-Atom Method. *Phys. Rev. B* 1989, *39* (11), 7441–7452.
29. Baskes, M. I. Modified Embedded-Atom Potentials for Cubic Materials and Impurities. *Phys. Rev. B* 1992, *46* (5), 2727–2742.
30. Jones, J. E. On the Determination of Molecular Fields—II. From the Equation of State of Gas. *Proc. Roy. Soc. A* 1924, *106* (738), 463–477.
31. Lennard-Jones, J. E. Wave Functions of Many-Electron Atoms. *Proc. Camb. Phil. Soc.* 1931, *27* (3), 469–480.
32. Zinenko, V. I.; Sorokin, B. P.; Turchin, P. P. *Fundamentals of Solid State Physics;* Izd-vo fiz.-mat. lit.:Moscow, 2001; p 335.

33. Stoddard, S. D.; Ford, J. Numerical Experiments on the Stochastic Behavior of a Lennard-Jones Gas System. *Phys. Rev. A* 1973, *8* (3), 1504–1512.
34. Krivtsov, A. M.; Krivtsova, N. V. Particle Method and Its Use in the Mechanics of a Deformable Solid. *Far-Eastern Math. J.* 2002, *3* (2), 254–276.
35. Wilson, N. T. The Structure and Dynamics of Noble Metal Clusters. PhD Thesis, University of Birmingham, September 2000.
36. Stockmayer, W. H. Second Virial Coefficient of Polar Gases. *J. Chem. Phys.* 1941, *9* (5), 398–402.
37. Kaplan, I. G. *Introduction to the Theory of Intermolecular Interactions;* Nauka: Main Edition of Physical and Mathematical Literature: Moscow, 1982; p 312.
38. Morse, P. M. Diatomic Molecules According to the Wave Mechanics. II. Vibrational Levels. *Phys. Rev.* 1929, *34*, 57–64.
39. Girifalco, L. A.; Weizer, V. G. Application of the Morse Potential Function to Cubic Metals. *Phys. Rev.* 1959, *114* (3), 687–690.
40. Magomedov, M. N. On the Surface Properties of Nanodiamonds. *Phys. Solid State* 2010, *52* (6), 1206–1214.
41. Hirschfelder, J.; Curtiss, C.; Byrd, R. *Molecular Theory of Gases and Liquids;* Publishing House of Foreign Literature: Moscow, 1961; p 931.
42. Belosludov, R. V.; Igumenov, I. K.; Belosludov, V. R.; Shpakov, V. P. Dynamical and Thermodynamical Properties of the Acetylacetones of Copper, Aluminium, Indium, and Rhodium. *Molecular Physics* 1994, *82* (1), 51–66.
43. Kitaigorodsky, A. I.; Mirskaya, K. B.; Tovbis, A. B. Energy Lattice of Crystalline Benzene in the Atomic-Atomic Approximation. *Crystallography* 1968, *13* (2), 225–231.
44. Buckingham, R. A.; Corner, J. Tables of Second Virial and Low-Pressure Joule-Thomson Coefficients for Intermolecular Potentials with Exponential Repulsion. *Proc. Roy. Soc. A* 1947, *189* (1016), 118–129.
45. Grechanovsky, A. E.; Eremin, N. N.; Urusov, V. S. Radiation Stability of $LaPO_4$ (Monazite Structure) and $YbPO_4$ (Zircon Structure) Based on Computer Simulation Data. *Solid State Phys.* 2013, *55* (9), 1813–1819.
46. Watanabe, T.; Ohdomari, I. Modeling of SiO_2/Si(100) Interface Structure by Using Extended-Stillinger-Weber Potential. *Thin Solid Films* 1999, *343–344*, 370–373.
47. Abell, G. C. Empirical Chemical Pseudopotential Theory of Molecular and Metallic Bonding. *Phys. Rev. B* 1985, *31* (10), 6184–6196.
48. Tersoff, J. New Empirical Model for the Structural Properties of Silicon. *Phys. Rev. Lett.* 1986, *56* (6), 632–635.
49. Ruda, M.; Farkas, D.; Abriata, J. Interatomic Potentials for Carbon Interstitials in Metals and Intermetallics. *Scripta Materialia* 2002, *46* (5), 349–355.
50. Tomar, V.; Zhou, M. Classical Molecular-Dynamics Potential for the Mechanical Strength of Nanocrystalline Composite fcc Al+α-Fe_2O_3. *Phys. Rev. B* 2006, *73* (17), 174116.1–174116.16.
51. Hohenberg, P.; Kohn, W. Inhomogeneous Electron Gas. *Phys. Rev. B* 1964, *136* (3), 864–871.
52. Baskes, M. I. Determination of Modified Embedded Atom Method Parameters for Nickel. *Mater. Chem. Phys.* 1997, *50* (2), 152–158.

53. Verzhbitsky, V. M. *Fundamentals of Numerical Methods: A Textbook for Universities;* 2nd ed., Revised; Vysshaja shkola: Moscow, 2005; p 840.
54. Samarsky, A. A.; Gulin, A. V. *Numerical Methods: A Textbook for High Schools;* Nauka: Moscow, 1989; p 432.
55. Morozov, A. I. *Solid State Physics. Crystal Structure. Phonons: A Study Guide;* MIREA: Moscow, 2006; p 151.
56. Suzdalev, I. P. *Nanotechnology: Physicochemistry of Nanoclusters, Nanostructures and Nanomaterials;* KomKniga: Moscow, 2006; p 592.
57. Cheng, A.; Merz, K. M. Application of the Nosé–Hoover Chain Algorithm to the Study of Protein Dynamics. *J. Phys. Chem.* 1996, *100* (5), 1927–1937.
58. Allen, M. P.; Tildesley, D. J. *Computer Simulation of Liquids;* Clarendon Press: Oxford, 2002; p 350.
59. Andersen, H. C. Molecular Dynamics Simulations at Constant Pressure and/or Temperature. *J. Chem. Phys.* 1980, *72* (4), 2384–2393.
60. Nose, S. A. Molecular Dynamics Method for Simulations in the Canonical Ensemble. *Molec. Phys.* 1984, *52* (2), 255–268.
61. Lemak, A. S.; Balabaev, N. K. A Comparison between Collisional Dynamics and Brownian Dynamics. *Mol. Simul.* 1995, *15* (4), 223–231.
62. Lemak, A. S.; Balabaev, N. K. Molecular Dynamics Simulation of Polymer Chain in Solution by Collisional Dynamics Method. *J. Comput. Chem.* 1996, *17* (15), 1685–1695.
63. Landau, L. D.; Teller, E. On the Theory of Sound Dispersion. *Physik. Zeits. Sowjetunion* 1936, *10* (1), 34–43.
64. Berendsen, H. J. C.; Postma, J. P. M.; van Gunsteren, W. F.; DiNola, A.; Haak, J. R. Molecular Dynamics with Coupling to an External Bath. *J. Chem. Phys.* 1984, *81* (8), 3684–3690.
65. Golo, V. L.; Shaitan, K. V. The Dynamic Attractor in the Berendsen Thermostat and the Slow Dynamics of Biomacromolecules. *Biophysics* 2002, *47* (4), 611–617.
66. Hoover, W. Canonical Dynamics: Equilibrium Phase-Space Distributions. *Phys. Rev. A* 1985, *31* (3), 1695–1697.
67. Hoover, W. G. Isomorphism Linking Smooth Particles and Embedded Atoms. *Physica A* 1998, *260* (3–4), 244–254.
68. Nose, S.; Klein, M. Constant Pressure Molecular Dynamics for Molecular Systems. *Mol. Phys.* 1983, *50* (5), 1055–1076.
69. Parrinello, M.; Rahman, A. Polymorphic Transitions in Single Crystals: A New Molecular Dynamics Approach. *J. Appl. Phys.* 1981, *52* (12), 7182–7190.

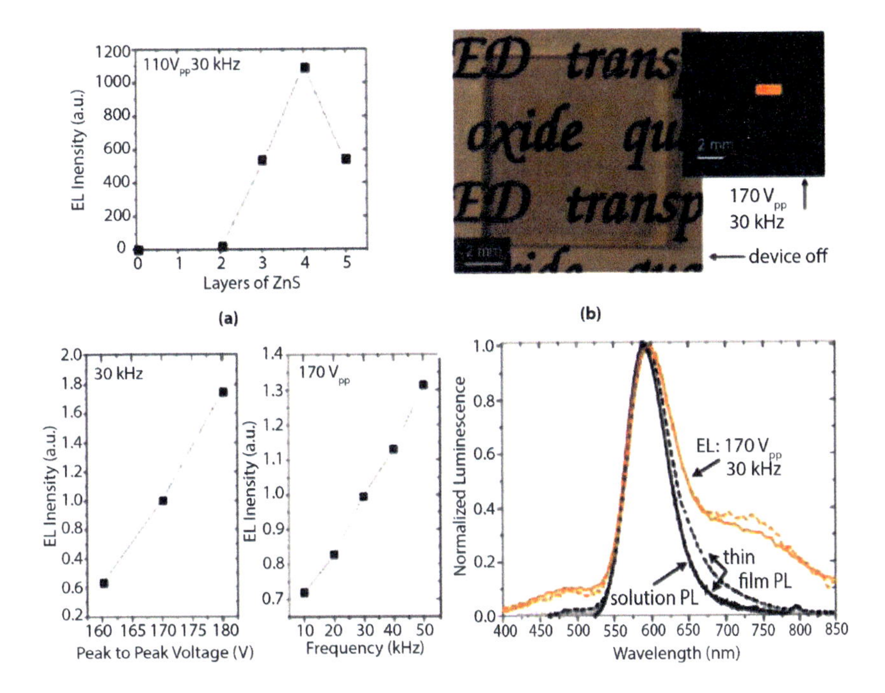

FIGURE 1.7 (a) Graph of dependence of electroluminescence intensity on the number of alternating layers of ZnS:Mn. (b) Photograph of ELS sample without the current supply and at electroluminescence excitation by the field with the frequency of 30 kHz and voltage of 170 V; (c) dependences of electroluminescence intensity on the voltage (at the left) and frequency (at the right) of the excitation electric field; (d) spectra of electroluminescence samples with four alternating layers of ZnS with dielectric layers of Al_2O_3 (solid orange line) and HfO_2 (dashed orange line) and luminescence spectra of ZnS:Mn particles in the form of suspension in chlorophorm (dashed black line) and in the form of film in the structure ITO/Al_2O_3/ZnS:Mn/30 (solid black line).
Source: Reprinted with permission from Ref. [20]. Copyright 2009 American Chemical Society.

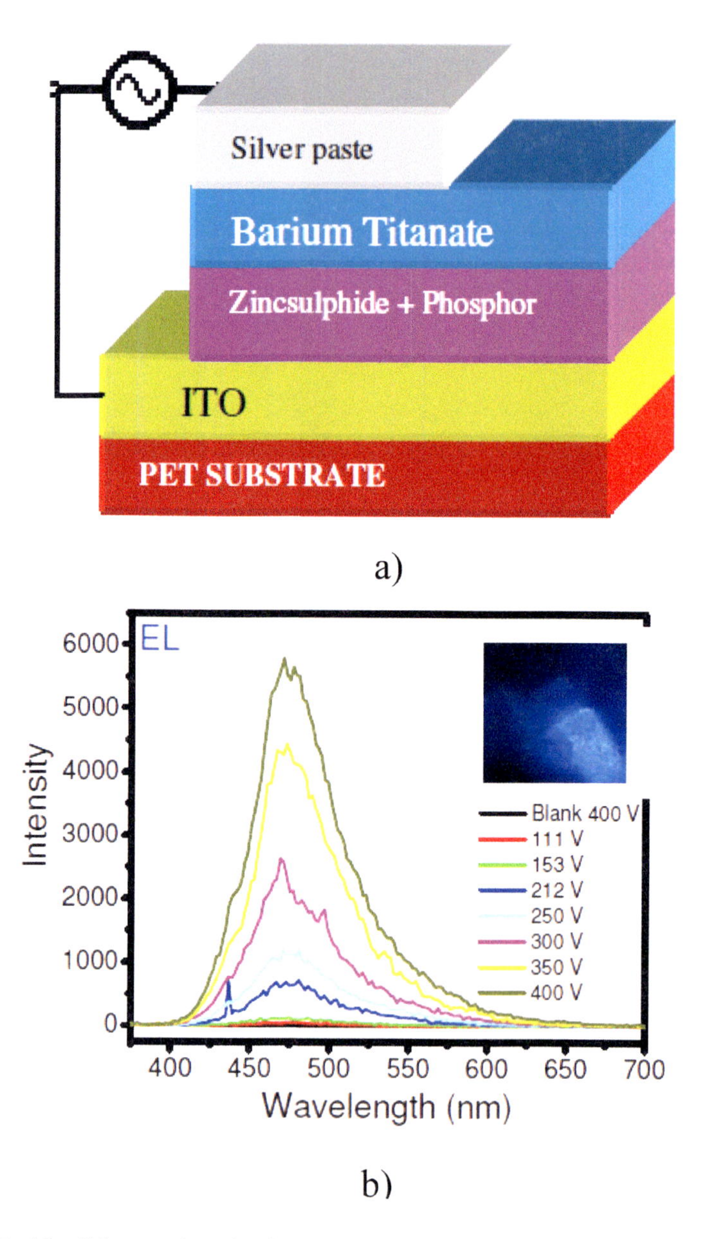

FIGURE 1.8 Scheme of ELS with working layers based on ZnS:Cu micro- and nanoparticles (a) and electroluminescence spectra of the sample with phosphor based on microparticles (b).

Sources: Reprinted with permission from Ref. [21]. Copyright 2015 AIP Publishing.

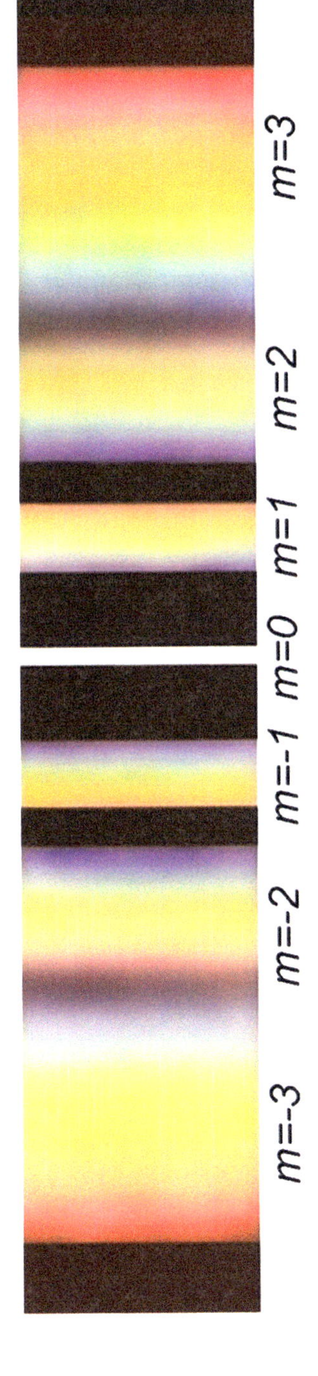

FIGURE 5.2 White light dispersion into the spectrum with the help of diffraction grid.

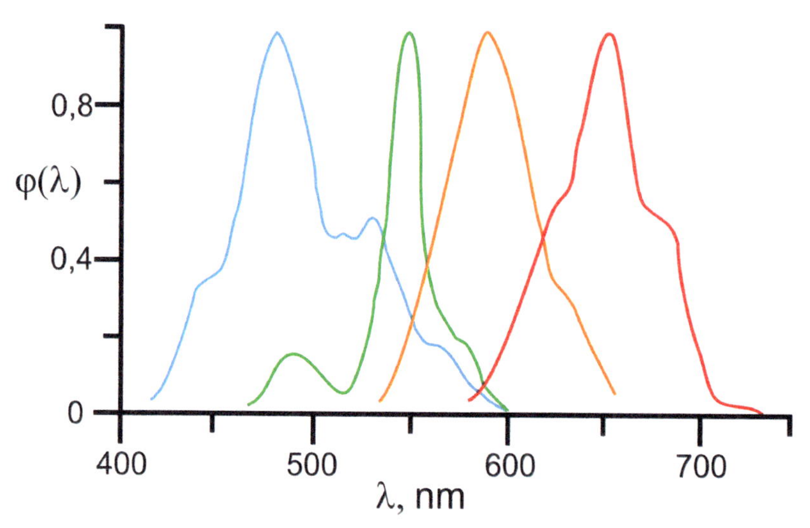

FIGURE 5.5 Spectral characteristics of electroluminescence (in relative units) of the films based on Ref. [9]: *SrS:Ce (blue line), ZnS:TbF$_3$ (green line), ZnS:Mn (orange line) and CaS:Eu,Ce (red line)*.

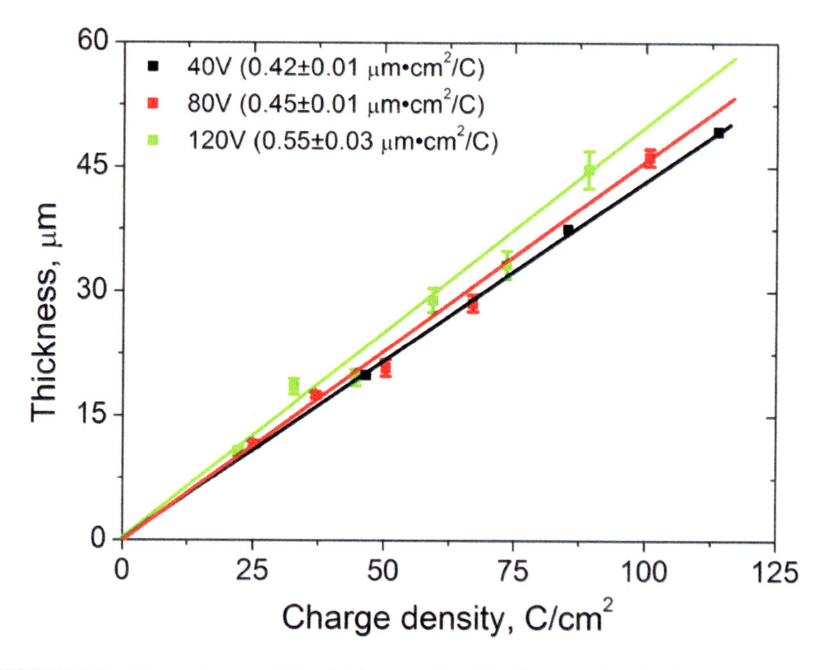

FIGURE 7.3 Dependence of the thickness of oxide films synthesized under different voltages on the density of the charge passed during anodization.

CHAPTER 7

Synthesis of Electroluminescent Nanostructures of ZnS Doped by Cu and Mn Ions

ABSTRACT

This chapter presents the results of experimental investigations of synthesis of electroluminescent nanostructures of ZnS alloyed by Cu and Mn ions. At first, the synthesis of AAO porous matrixes is described. The results of coulometric control of oxide film thickness and pore diameter control are given. Then, the results of alloyed ZnS deposition are presented. In this part of the chapter, the facility of ultrahigh-vacuum evaporation based on the vacuum system USU-4 are described and results of experimental investigations of deposition of ZnS alloyed with Cu and Mn onto smooth templates and matrixes of porous AOA are presented and discussed.

7.1 SYNTHESIS OF AAO POROUS MATRIXES

The obtaining of materials with certain physical and chemical properties is one of the most important stages on the way to produce the operating semiconductor device. The methods of semiconductor technology are mainly applied in production of thin-film electroluminescent light sources, which allow providing high uniformity of structure and properties over a large area, and this is rather a complicated technological problem. This is especially related to the methods based on the processes of thermal evaporation of materials. But it should be pointed out that in comparison with chemical methods, such as from chemical vapors deposition and electrochemical deposition, methods of thermal deposition allow obtaining purer films (there is no contamination with substance residues used in chemical reactions), and in situ control of the structure and composition of the deposited material allows correcting the deposition modes. Besides, the processes taking place during the film formation on the surface of smooth templates, regardless of their obtaining methods, were studied in detail long time ago.

The material deposition onto the templates with well-developed surface is a more complicated process not completely studied. Thus, the evaporation processes with the formation of material gaseous phase are similar to the deposition onto the template smooth surface, but it is still not clear how and at what temperature modes of the template the material is deposited into the pores, e.g., of anodic aluminum oxide films.

The thing is that the cosinusoidal law is the common distribution law during the propagation of material gaseous beam. When deposited onto a smooth surface, all atoms, which reached the template, can condense.

When being deposited onto developed porous surfaces, atoms of the material deposited can either freely penetrate to the pore bottom, or deposit on its walls, thus leading to its fast dusting and the pore will not be completely filled. In this case, it is necessary to carry out investigations on the adhesion properties of the material deposited depending on the template temperature modes, characteristics of the porous structure of the matrix being filled (pore diameters and distance between them, thickness of the porous film).

The following reagents and materials were used to synthesize porous matrixes of anodic aluminum oxide: Al (0.5-mm thick foil, purity—99.999%), oxalic acid $(COOH)_2$ (Aldrich, 98%), phosphorous acid $H_3PO_{4\,(conc.)}$ (especially pure), CrO_3 (chemically pure). All water solutions were prepared with distilled water.

Aluminum foil pre-polished electrochemically in the mixture of 185 g/L of chromium anhydride (CrO_3) and 1480 g/L of orthophosphoric acid (H_3PO_4) at 80°C was used as the initial material for the synthesis of porous matrixes.

The galvanostatic pulse mode of polishing with electric current density of 650 mA/cm^2 was applied. The pulse duration was 5 s, interval between pulses was 40 s, and voltage constraint in the cycling process was 20 V. The duration of electrochemical polishing cycle was 40 pulses.

The anodic oxidation of aluminum was carried out in two-electrode electrochemical cell (Fig. 7.1) with the use of DC power supply. Here, the stainless steel ring served as the cathode, and aluminum foil as the anode.

The anodic oxidation was carried out with constant pumping of the electrolyte (0.3 M $H_2C_2O_4$), preliminarily cooled down in thermostat to 0°C. The dependence of current force upon time was registered during anodizing with the help of digital multi-meter.

The ordered porous matrixes were formed by two-stage anodizing under 40 V. When applying this technique, the oxide film formed after the first anodizing was completely diluted in the mixture of 20 g/L of CrO_3 and 35 mL/L< of H_3PO_4 at 60°C.

As a result, the structure of cavities ordered into the hexagonal array was formed on aluminum surface. The structured surface of aluminum was oxidized again under the same conditions.

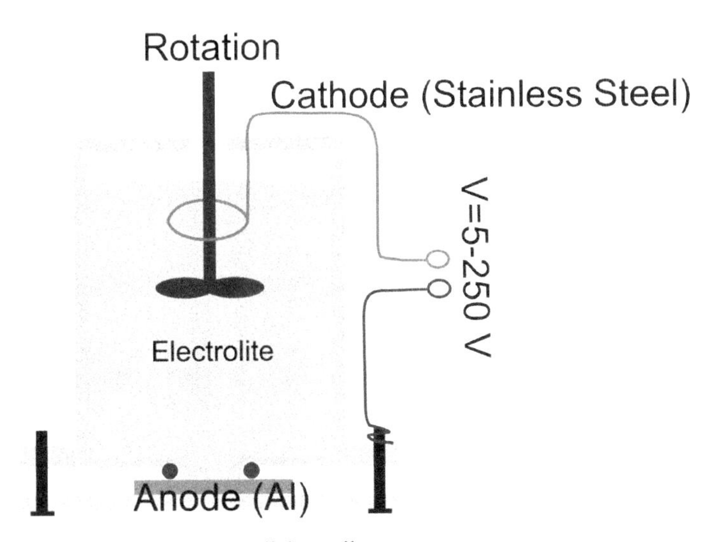

FIGURE 7.1 Scheme of the anodizing cell.

To obtain the oxide matrixes with the specified pore diameter, after anodizing, the porous structure was additionally etched in the solution of 5 mass % of H_3PO_4 at 25°C during different time intervals.

7.1.1 Coulometric Control of Oxide Film Thickness

Since the process of oxide film formation is electrochemical, the mass of the oxide film formed and, consequently, its thickness are determined by the charge during anodization. Therefore, in the frameworks of this work, the method of coulometric control was used for the precision control of the oxide film thickness. Chronoamperometric dependencies obtained in the modes of "soft" anodization (with current density under 10 mA/cm^2) and in the mode of "hard" anodization (with current density over 10 mA/cm^2) are given in Figures 7.2a and b, respectively.

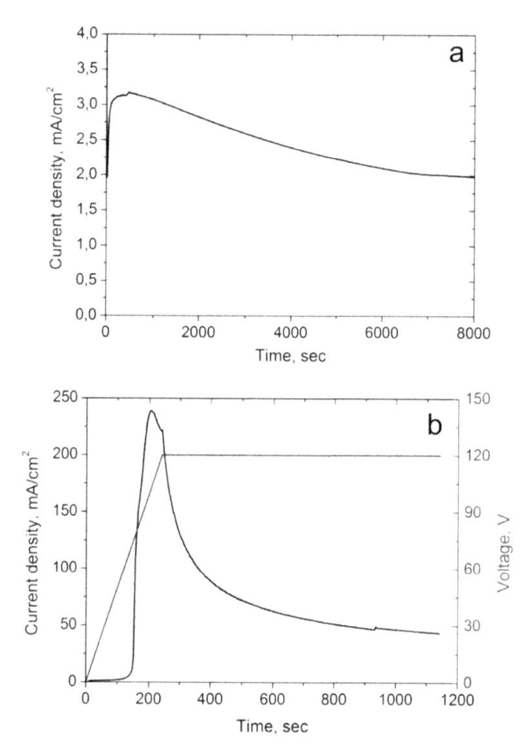

FIGURE 7.2 Chronoamperometric dependencies obtained during aluminum anodic oxidation under the voltages of 40 V (a) and 120 V (b).

While applying the "hard" anodization mode, at the first stage, the voltage increases linearly with the rate of 0.5 V/sto avoid the dielectric breakdown of the oxide film. To determine the coefficient binding the density of the charge, passed during anodization, with the oxide film thickness, in this work, the films of anodic aluminum oxide with different thicknesses were synthesized by passing the specified charge under the voltages of 40, 80, and 120 V.

The thickness of the film formed was found with the help of micrometer and by the microphotograph of the oxide film chipping. The graph of dependence of the thickness of the oxide film formed on the density of the charge passed during anodization is demonstrated in Figure 7.3.

The oxide film thickness is proportional to the charge passed during anodization, at the same time, the proportionality coefficient increases with the anodization voltage growth from 0.415 ± 0.005 (for the voltage of 40 V) up to 0.550 ± 0.030 (μm·cm²)/C (for the voltage of 120 V).

The change in the proportionality coefficient linking the charge passed and oxide film thickness can be connected with different rates of side reactions, thus decreasing the current output when the anodization voltage is diminished. Besides, the change in the film porosity or aluminum oxide density formed under different voltages can be the possible reasons of the coefficient variation.

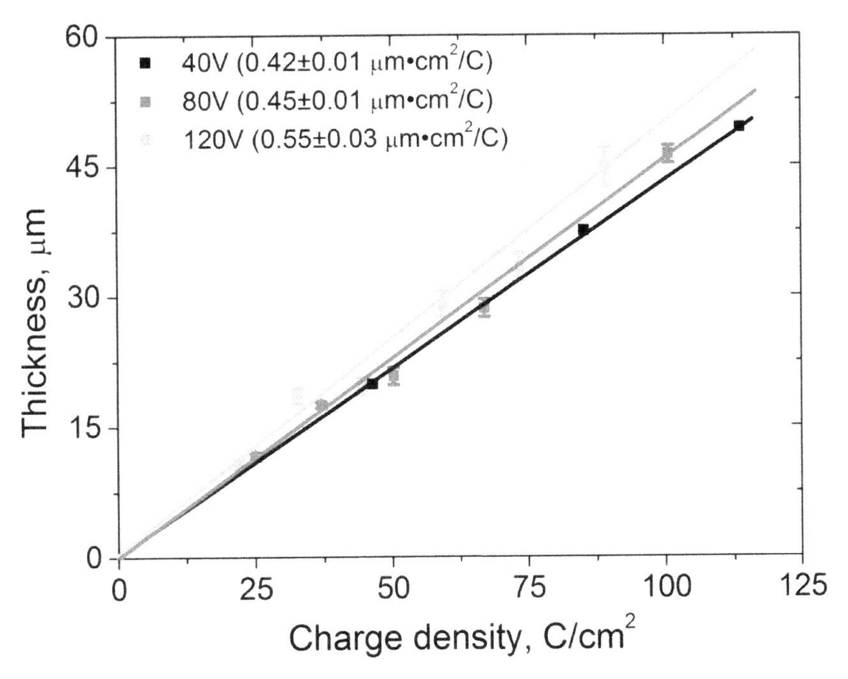

FIGURE 7.3 **(See color insert.)** Dependence of the thickness of oxide films synthesized under different voltages on the density of the charge passed during anodization.

7.1.2 Pore Diameter Control

The pore diameter of anodic aluminum oxide can vary because of the changes in anodization voltage; however, when the anodizing voltage increases, the oxide film porosity is preserved approximately equaling 10%. To increase the oxide matrix porosity, and, consequently, to raise the filling degree of the oxide matrix for increasing the pore diameter, the technique of chemical etching in 5 mass % of phosphoric acid was applied in this work.

The oxide films synthesized by anodic oxidation under “soft” conditions and voltage of 40V were etched during different time intervals. The average pore diameter and pore dispersion by sizes on the external surface of the oxide film were found by the statistic processing of microphotographs in software ImageJ.[1]

The microphotographs of the initial oxide film, as well as the film obtained by etching in acid solution within 15, 30, 45, 60, and 75 min are given in Figure 7.4.

The dependence of the average pore diameter upon etching duration is demonstrated in Figure 7.5. On the basis of the analysis of the average pore diameter dependence on etching duration, it was found that the pore etching rate is 0.54 nm/min.

The etching proceeds until the average pore diameter of about 80 nm; further etching results in the destruction of honeycomb skeleton consisting of denser aluminum oxide and complete porous structure destruction.

Porous films with large pore diameters can be obtained by ordinary chemical etching when applying the structure formed under “soft” conditions as the initial film. In this case, the maximum pore diameter is limited within 70–80 nm. To form the matrixes with larger channel diameters, it is necessary to use the oxide films formed in the mode of “hard” anodization.

However, when using such films, we come across the problem of small distance between the pore centers on the upper surface of the oxide film (Fig. 7.6) that can be explained both by the random pore generation on the metal initial surface and linear time base by voltage at the initial stage of anodization.

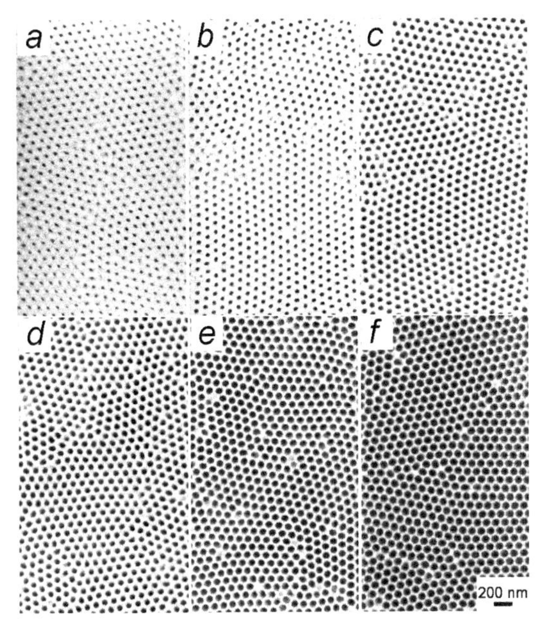

FIGURE 7.4 SEM images of anodic aluminum oxide film etched during (a) 0 min, (b) 15 min, (c) 30 min, (d) 45 min, (e) 60 min, and (f) 75 min.

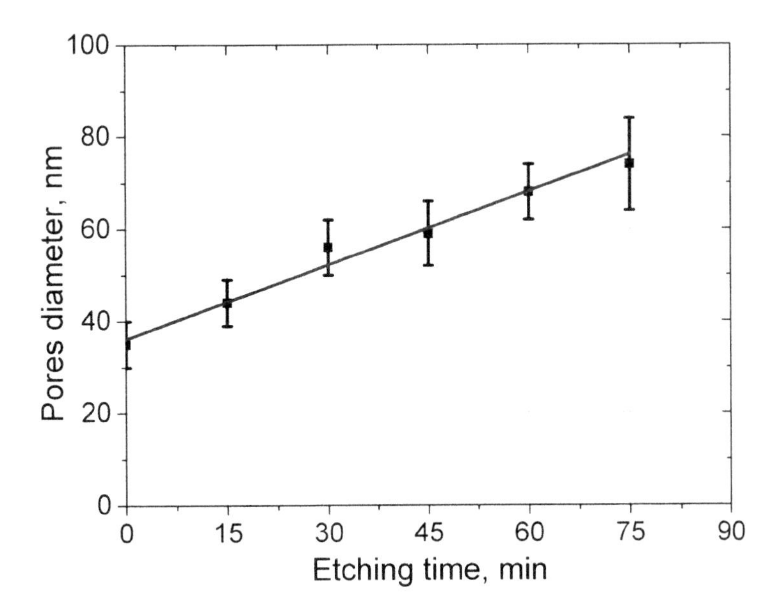

FIGURE 7.5 Dependence of the average pore diameter on oxide film etching duration.

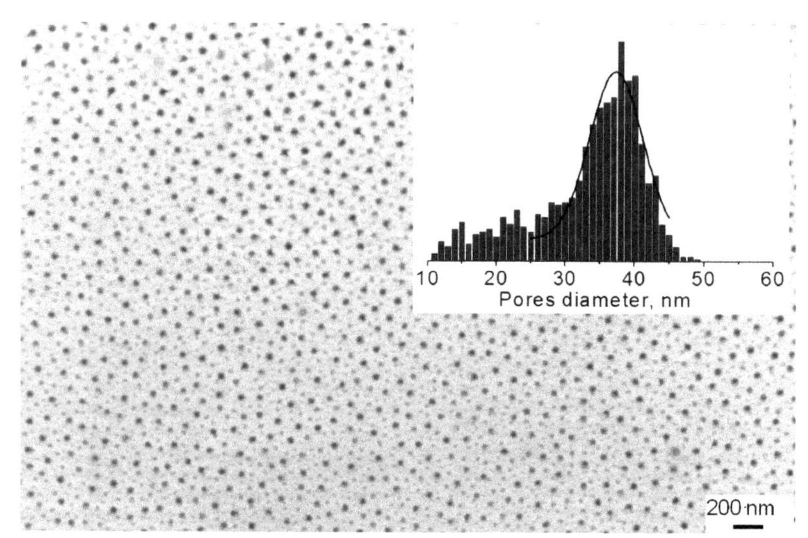

FIGURE 7.6 Microphotograph of the upper surface of the oxide film synthesized by the method of one-stage anodization under the voltage of 120 V.

It should be pointed out that the problem of pore random generation at the initial stage of anodization is solved quite easily applying the two-stage anodization method. At the same time, it is rather problematic to change the mode from the linear time base to the initial inclusion of the required voltage value (80 V or 120 V), since we come across the problem of dielectric breakdown of oxide film. Therefore, we tried to use the intermediary stage, namely, the protective layer formation by the anodic oxidation of metal with the structured surface after the removal of the oxide film formed during the first stage of anodization to avoid the dielectric breakdown of oxide film when forming the porous film at the next stage of anodization. To form the protective layer, after the first anodization and oxide film removal in the mixture of 20 g/L of CrO_3 and 35 mL/L of H_3PO_4 at 60°C, the metal was subjected to anodic anodization in 0.1M of H_3PO_4 at 5°C and voltage 10–15V greater than the voltage applied to form the porous structure. The chronoamperometric dependence obtained during the protective layer formation is shown in Figure 7.7a. The current density decrease after the voltage reaches its stationary value indicates the formation of the barrier-type oxide layer. During the next anodization (chronoamperogram is given in Fig. 7.7b), practically zero current density is observed within 60 s, thus corresponding to the process of chemical dissolution of the oxide film. After the protective layer is dissolved till the thickness sufficient for the voltage used during the oxide film formation to be capable of transporting ions thorough the oxide layer, the current density increases up to 300 mA/cm^2 that corresponds to the oxide film formation under "hard" conditions of anodization. The matrix can be additionally etched in phosphoric acid after the anodization to increase the channel diameter following the technique described before. The micro-photographs of the surfaces of oxide films formed with the use of 3-stage technique comprising the intermediary stage of protective layer formation by anodization in 0.1M of H_3PO_4 and voltages of 80 and 120 V are given in Figures 7.8a and 7.9a. After etching these films in 5 mass % of H_3PO_4 within 75 and 105 min for the voltages of 80 and 120 V, respectively, it is possible to form the oxide matrixes with the average pore diameters of 90 ± 30 and 170 ± 28 nm, respectively. The microphotographs of the oxide films after pore etching are demonstrated in Figures 7.8b and 7.9b.

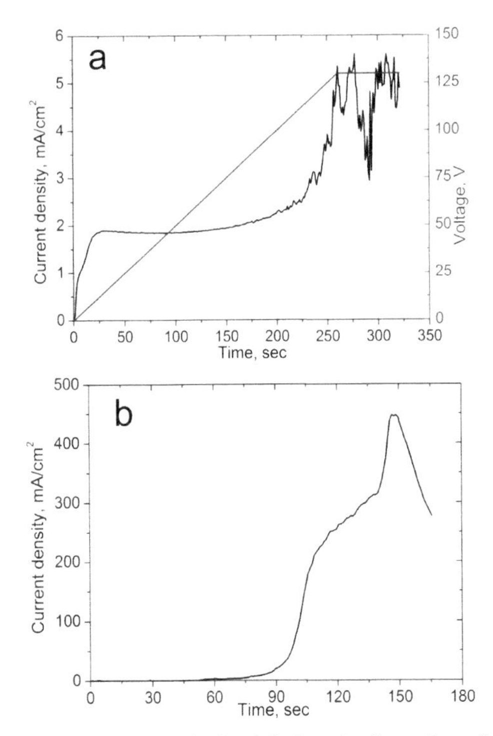

FIGURE 7.7 Chronoamperograms obtained during the formation of (a) protective layer in 0.1M of H_3PO_4 and (b) during the formation of the porous structure under the voltage of 120 V in 0.3M $H_2C_2O_4$.

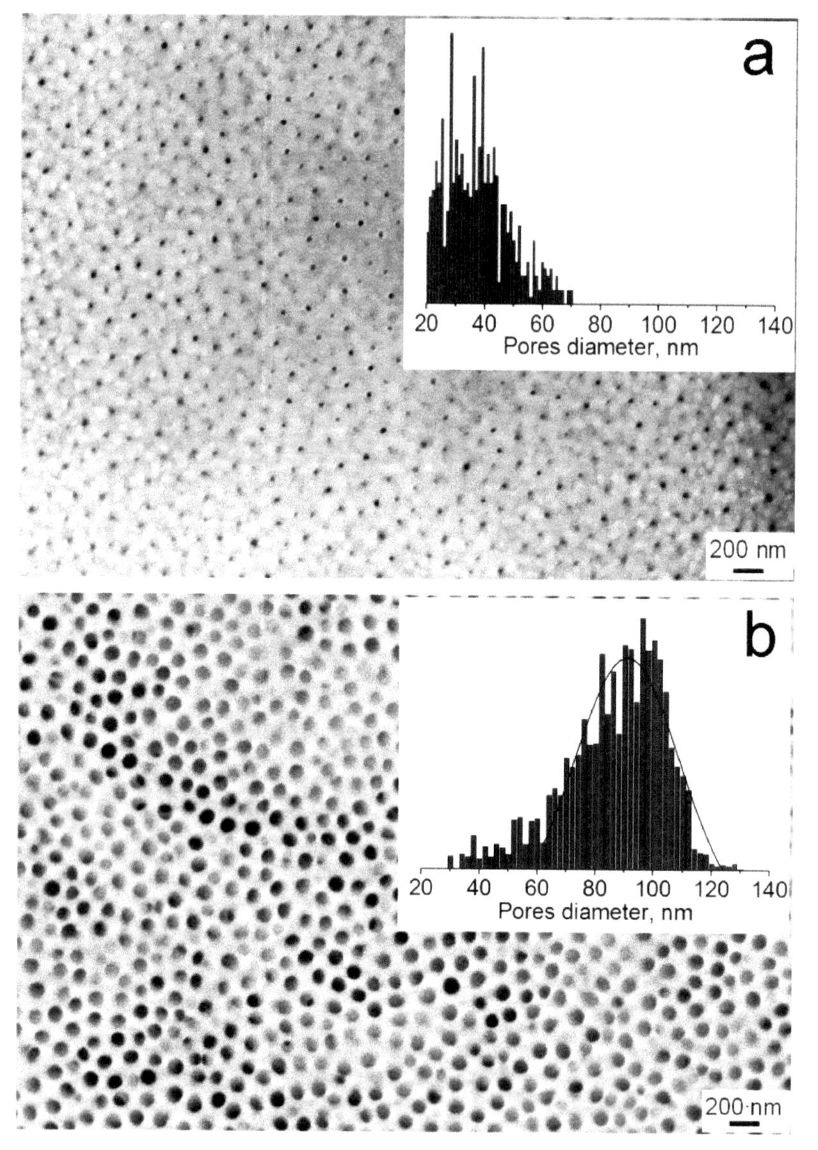

FIGURE 7.8 Microphotograph of the porous film of aluminum oxide formed under the voltage of 80 V in 0.3M of $H_2C_2O_4$ with the intermediary stage of protective layer formation in 0.1M of H_3PO_4 under the voltage of 85 V. (a) Initial film and (b) film after etching in 5 mass % of H_3PO_4 within 75 min.

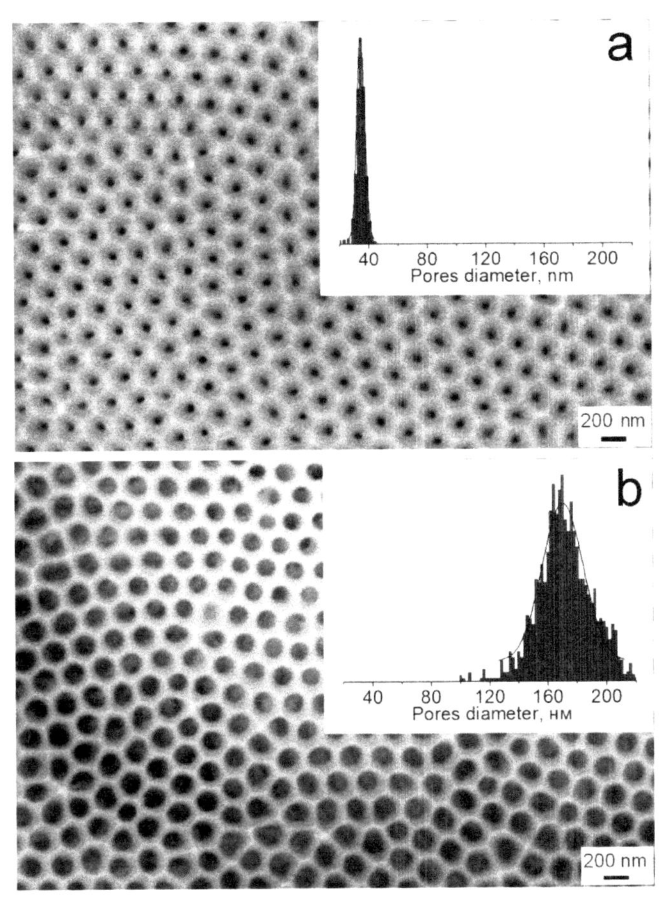

FIGURE 7.9 Microphotograph of the porous film of aluminum oxide formed under the voltage of 120 V in 0.3 M of $H_2C_2O_4$ with the intermediary stage of protective layer formation in 0.1M of H_3PO_4 under the voltage of 130 V. (a) Initial film and (b) film after etching in 5 mass % of H_3PO_4 within 105 min.

Thus, the technique, which allows controlling both the pore diameter and porous matrix thickness, is proposed. Besides, the etching technique allows increasing the porosity and, consequently, achieving the maximum degree of pore filling afterwards.

Twenty-four matrixes of porous anodic aluminum oxide with the pore diameters of 40 and 80 nm and with and without aluminum substrate having minimum distances between the pores not exceeding 20 nm were prepared for further investigations.

7.2 DEPOSITION OF DOPED ZINC SULFIDE

7.2.1 Facility of Ultrahigh-Vacuum Evaporation Based on the Vacuum System USU-4

The facility of ultrahigh-vacuum evaporation based on the vacuum system USU-4 was applied to deposit ZnS doped with Cu and Mn (Fig. 7.10). The facility consists of three main units.

FIGURE 7.10 Facility of ultrahigh-vacuum evaporation of semiconductors based on the vacuum system USU-4.

(1) Template-loading chamber (Fig. 7.11) with independent system of vacuum pumping based on spiral pre-vacuum pump Aglient SH-110 (Netherlands) and turbomolecular pump NMT-100 (Iskitim, Russia). The chamber pumping is fully automated, that is, after the required pressure of 5 Pa is achieved at the outlet of the pre-vacuum pump, the turbomolecular pump is automatically started, which provides the pressure in the loading chamber of no worse than 10^{-5} Pa. The same pumping means are applied to produce pre-vacuum in two other chambers of the facility. The pressure is controlled in the chamber with the help of Penning-type sensor PMM-32-1 and vacuum gauge AV3592 ("Avaks," Russia). The templates

are loaded through the sluice vacuum window; the template holder is fixed on the stock of magnetic manipulator with the help of the threaded joint.

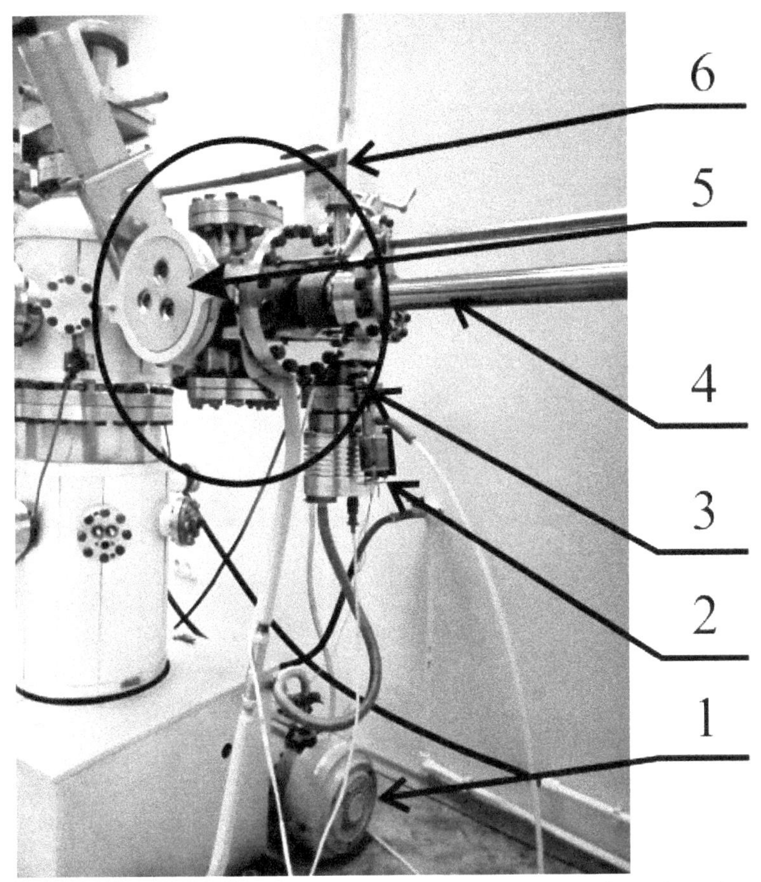

FIGURE 7.11 Chamber of loading and preliminary preparation of templates. 1—spiral fore-vacuum pump Aglient SH-110, 2—turbomolecular pump NMT-100, 3—main chamber, 4—magnetic manipulator, 5—sluice device, and 6—pumping pipeline from the deposition chamber.

(2) Analytical chamber USU-4 (Fig. 7.12) with independent pumping with the help of ion pump NMD-0,4. At present, it is used as the intermediary chamber between the template-loading chamber and evaporation chamber. It is planned to install Auger electron spectrometer into it later to carry out the in situ analysis of the chemical composition of the samples

obtained. The Penning-type pump in this chamber is also used to achieve the operating vacuum in the evaporation chamber.

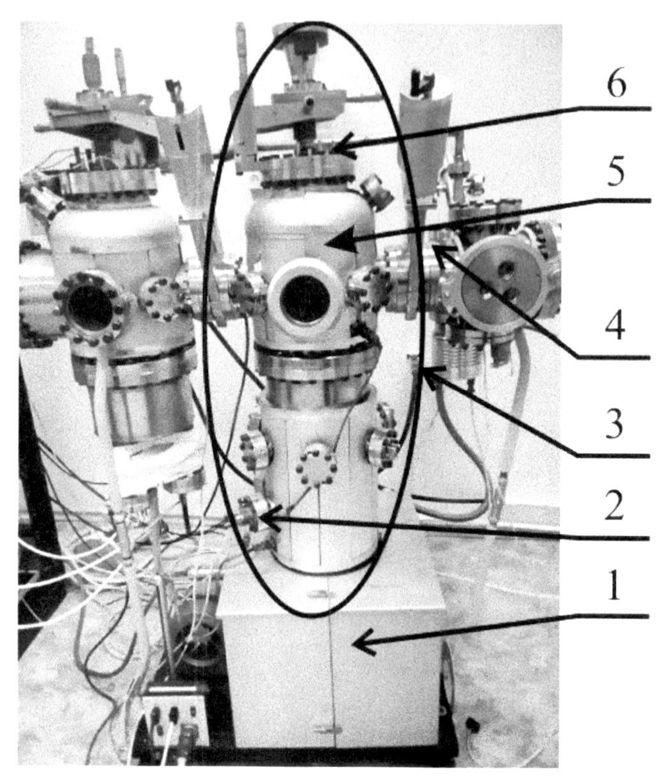

FIGURE 7.12 Analytical chamber. 1—ion pump NMD-0,4, 2—input of titanium sublimation pump, 3—main chamber, 4—gate between the loading chamber and analytical chamber, 5—analytical chamber, and 6—rotation device.

(3) Chamber of ultrahigh-vacuum (no worse than 10^{-7} Pa) evaporation of semiconductors based on ZnS (Fig. 7.13). To maintain pressure during the evaporation, the chamber is equipped with ion pump RIBER. The system of vacuum pipelines allows carrying out pre-vacuum and ultrahigh-vacuum chamber pumping, bypassing the main chamber USU-4. The vacuum state is controlled with Penning-type sensor PMM-32-1 and vacuum gauge AV3532 ("Avaks," Russia). The deposited material is evaporated from the indirect-heating Knudsen evaporation cells (of MBE type) with the crucibles of pyrolytic boron nitride (Fig. 7.14).

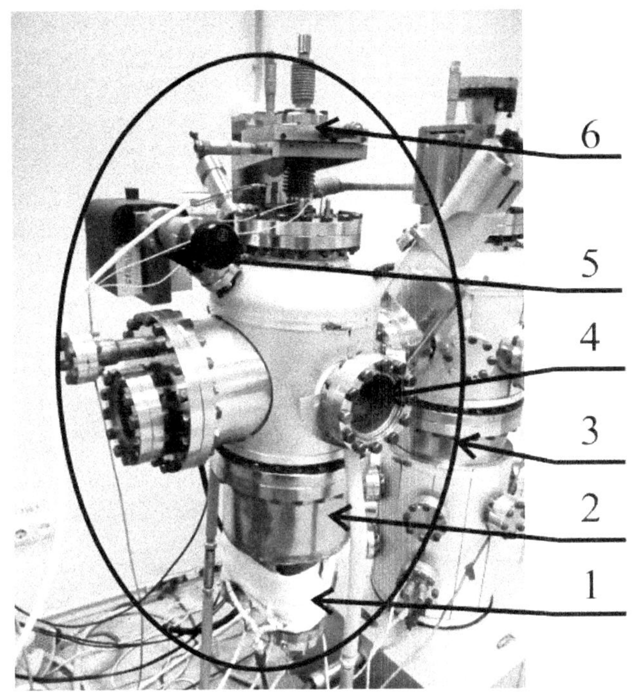

FIGURE 7.13 Chamber of semiconductor evaporation. 1—evaporator unit, 2—cooling jacket of evaporator unit, 3—main chamber, 4—window to observe the evaporation process, 5—ion pump Riber, and 6—template rotation device.

FIGURE 7.14 Indirect-heating Knudsen evaporation cell (of MBE type) with the crucible of pyrolytic boron nitride.

All in all, the chamber comprises three cells for the evaporation of ZnS, Cu, and MnS. The crucibles with the material can be replaced after the chamber opening to the atmosphere. The material for evaporation is used in the form of powder, thus providing more intensive sublimation and, consequently, rather high rate of film and nanostructure growth on the templates. Each cell is equipped with the gate and thermocouple for accurate measuring of crucibles temperature and maintaining their constant temperature. The heating is performed by supplying current to the heating element from the power sources ATTEN TPR6010S controlled by the software and hardware complex based on the eight-channel programmable logic controller PLC-154 (OVEN, Russia). The modes of evaporation processes are controlled by the software developed in the visualization medium CoDeSys. The main window is demonstrated in Figure 7.15. The software interface allows establishing and maintaining heating temperatures of crucibles and templates, controlling the evaporator gates, and controlling the evaporation time.

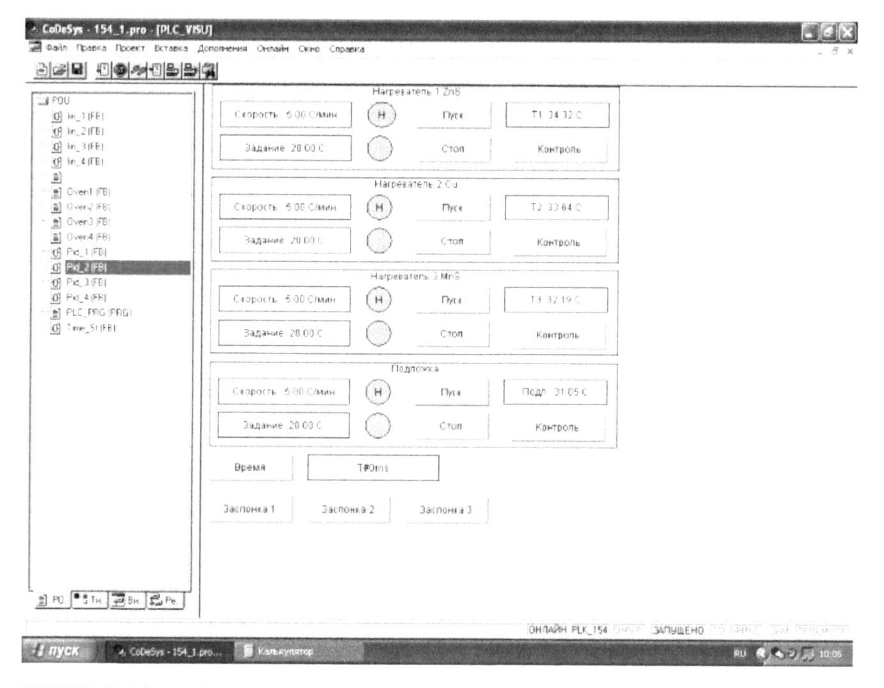

FIGURE 7.15 Main window of the software controlling evaporation processes.

Interchamber accessories (Fig. 7.16) represent the rotatable disc of stainless steel located at the distance of 20 cm from the upper edge of the crucibles with four slots for template holders (Fig. 7.17). Up to four templates were installed onto each template holder. So, it is possible to simultaneously deposit 16 samples. The disc with the heater and temperature sensor in the form of thermal resistance TSP-100 on its surface is installed above the template holder. The thermal resistance is connected to the controller PLC-154. The sensor has been precalibrated to the temperature readings of the disc with template holders, which are positioned so that during the disc rotation the trajectories of their centers are passing precisely above the crucibles. Thus, during the deposition, the template holder with samples is periodically above the crucible with the corresponding material evaporated.

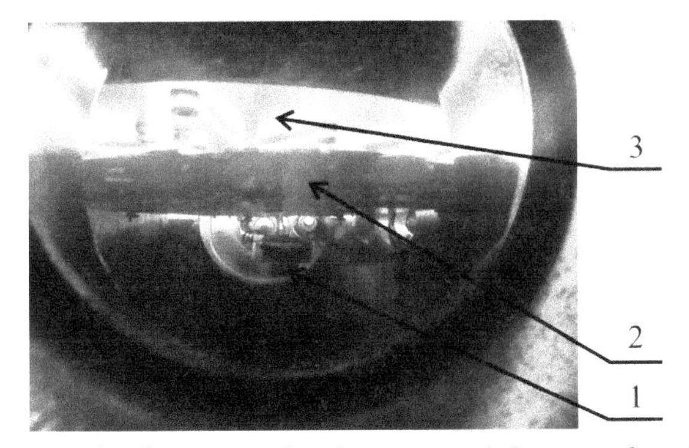

FIGURE 7.16 Interchamber accessories. 1—water-cooled sensor of control-measuring points (CMP), 2—rotatable disc of template holders, and 3—disc with the template heating spiral.

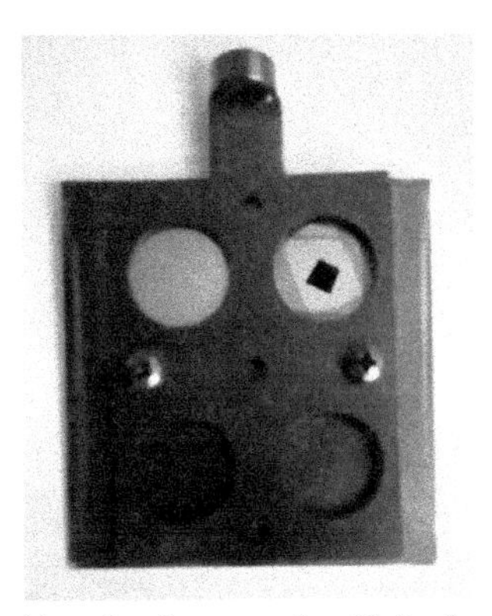

FIGURE 7.17 Template holders for four samples. Hole diameter is 16 mm.

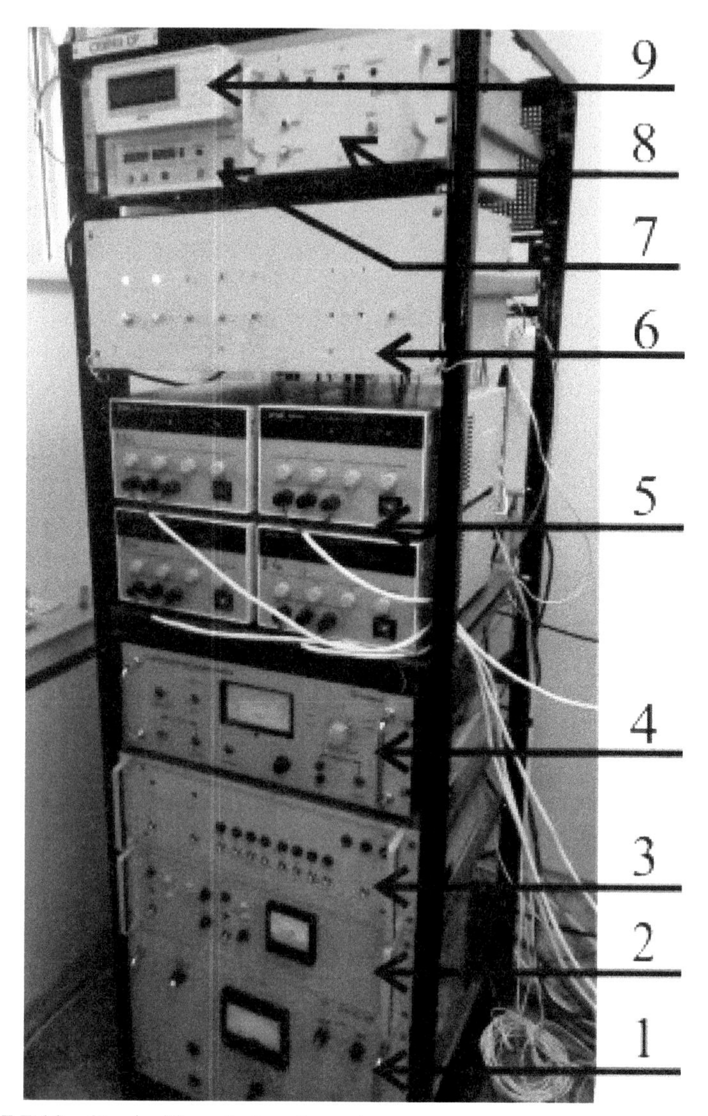

FIGURE 7.18 Stand with control equipment.
1—power supply unit of the ion pump NMD-0,4, 2—power supply unit of sublimation pump, 3—control unit of chamber wall annealing, 4—power supply unit of the ion pump Riber, 5—power supply units of crucible and template heating, 6—control unit of crucible and template heating, gates and pumping-out system, 7—vacuum gauge AV3692, 8—power supply unit of turbomolecular pump, and 9—control unit of CMP.

The maximum heating temperature of the crucibles is 1400°C, templates—400°C. The amount of the evaporated material and evaporation rate are measured with the help of water-cooled quartz crystal monitor of QCM ("Rosakadempribor," Russia). The analytical stand with the equipment is demonstrated in Figure 7.18.

7.2.2 Deposition of ZnS Doped with Cu and Mn onto Smooth Templates and Matrixes of Porous AOA

Simultaneous evaporation from at least two evaporators is necessary to deposit the impurity-doped ZnS. The powder sublimated at the crucible temperatures from 800°C up to 1000°C serves as sulfide zinc source, copper source—a piece of copper granule with 99.99% purity evaporated at 900–950°C, manganese source–manganese sulfide sublimated at 1200–1250°C. The rates of material deposition, whose dependencies are given in Figure 7.19, depend on the temperature. It is seen that the exponential increase of the film growth rate takes place with the temperature rise; therefore, when ZnS with the impurity is deposited, such crucible temperatures were set as to provide the evaporation rate, which results in the obtaining of the material with the required concentration of the impurity (5 and 10 at.%).

It should be pointed out that as the amount of the material evaporated in the crucible goes down, the temperature at which the evaporation starts decreases. Therefore, the crucible temperature was adjusted to have the constant deposition rate during evaporation, based on QCM readings. For example, when the deposition rate is 75 Å/min, the film about 4500 Å thick is growing during 1 h. Higher evaporation rates and, consequently, the crucible temperature result in ZnS "splitting" from the crucible, and QCM readings do not correspond to the reality.

The deposition was preformed onto the templates from glass, glass with ITO layer evaporated, silicon, polished aluminum, and matrix of porous aluminum oxide with different porous structures. The samples on the glass template were used to determine the thickness of the layers obtained by the method of optical spectroscopy, those on ITO and aluminum—as reference samples for electroluminescent investigations. Two types of matrixes of porous aluminum oxide were used: without aluminum base removal and with removed aluminum (Chapter 2).

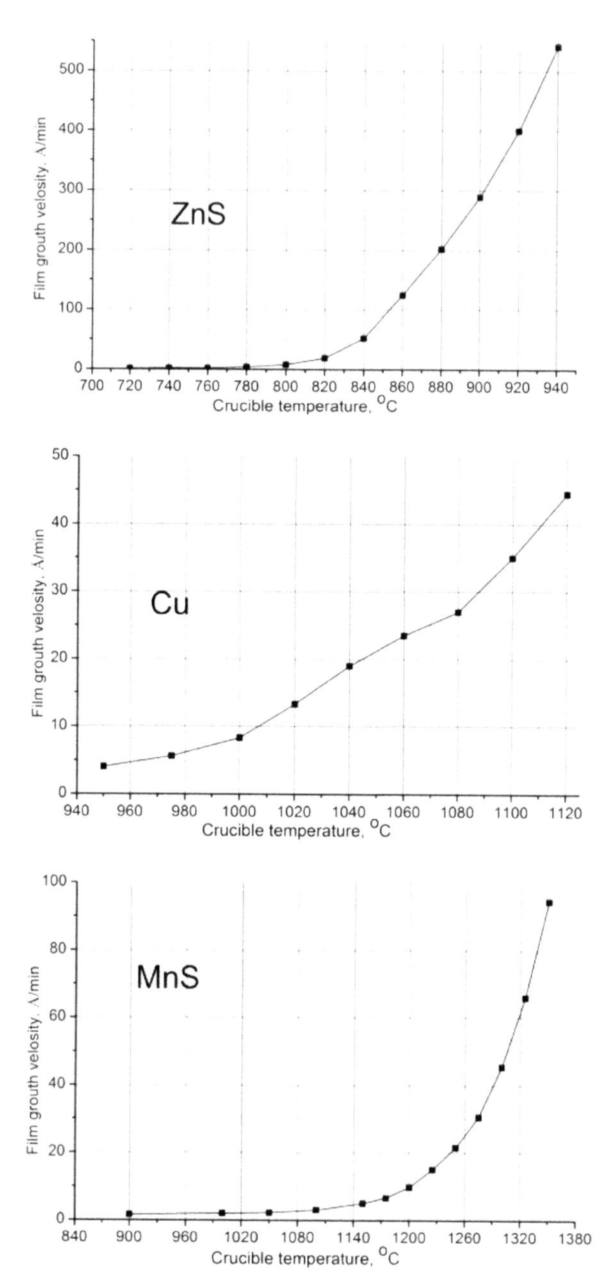

FIGURE 7.19 Dependencies of ZnS, MnS, and Cu film growth rate on the crucible temperature based on QCM data.

The list, nomenclature, and amount of the templates used are given in Table 7.1.

TABLE 7.1 The List, Nomenclature, and Amount of the Templates Used.

Template type	Designation	Amount
Cover glass	SiO_2	6
Cover glass with evaporated ITO layer	SiO_2/ITO	6
Al_2O_3 matrix without aluminum removal (d_{pore} = 40 nm)	AAO/Al_40	6
Al_2O_3 matrix with removed aluminum (d_{pore} = 40 nm)	AAO_40	6
Al_2O_3 matrix without aluminum removal (d_{pore} = 80 nm)	AAO/Al_80	6
Al_2O_3 matrix with removed aluminum (d_{pore} = 80 nm)	AAO_80	6

Deposition of doped ZnS was performed under the growth rate of 100 Å/min. The constant growth rate was adjusted by the temperature of the ZnS evaporator in automatic mode with the help of QCM. Copper doping was carried out at Cu evaporator temperatures of 900°C and 930°C, manganese at MnS evaporator temperatures of 1200°C and 1230°C.

The list and nomenclature of the samples obtained are given in Table 7.2.

TABLE 7.2 The List and Nomenclature of the Samples Obtained.

No.	Sample	Designation	Doping atom	T_{ev}, °C
1	ZnS, doped by Cu at 900°C on glass	Cu_900_SiO_2	Cu	900
2	ZnS, doped by Cu at 950°C on glass	Cu_930_SiO_2	Cu	930
3	ZnS, doped by Mn at 1250°C on glass	Mn_1200_SiO_2	Mn	1200
4	ZnS, doped by Mn at 1350°C on glass	Mn_1230_SiO_2	Mn	1230
5	ZnS, doped by Cu at 900°C on ITO/glass	Cu_900_SiO_2_ITO	Cu	900

TABLE 7.2 *(Continued)*

No.	Sample	Designation	Doping atom	T_{ev}, °C
6	ZnS, doped by Cu at 950°C on ITO/glass	Cu_930_SiO$_2$_ITO	Cu	930
7	ZnS, doped by Mn at 1250°C on ITO/glass	Mn_1200_SiO$_2$_ITO	Mn	1200
8	ZnS, doped by Mn at 1350°C on ITO/glass	Mn_1230_SiO$_2$_ITO	Mn	1230
9	ZnS, doped by Cu at 900°C on Al_2O_3 matrix without Al etching (d_{pore}=40 nm)	Cu_900@AAO/Al_40	Cu	900
10	ZnS, doped by Cu at 950°C on Al_2O_3 matrix without Al etching (d_{pore}=40 nm)	Cu_930@AAO/Al_40	Cu	930
11	ZnS, doped by Mn at 1250°C on Al_2O_3 matrix without Al etching (d_{pore}=40 nm)	Mn_1200@AAO/Al_40	Mn	1200
12	ZnS, doped by Mn at 1350°C on matrix without Al etching (d_{pore}=40 nm)	Mn_1230@AAO/Al_40	Mn	1230
13	ZnS, doped by Cu at 900°C on Al_2O_3 matrix with etched Al (d_{pore}=40 nm)	Cu_900@AAO_40	Cu	900
14	ZnS, doped by Cu at 950°C on Al_2O_3 matrix with etched Al (d_{pore}=40 nm)	Cu_930@AAO_40	Cu	930
15	ZnS, doped by Mn at 1250°C on Al_2O_3 matrix with etched Al (d_{pore}=40 nm)	Mn_1200@AAO_40	Mn	1200
16	ZnS, doped by Mn at 1350°C on Al_2O_3 matrix with etched Al (d_{pore}=40 nm)	Mn_1230@AAO_40	Mn	1230
17	ZnS, doped by Cu at 900°C on matrix without Al etching (d_{pore}=80 nm)	Cu_900@AAO/Al_80	Cu	900
18	ZnS, doped by Cu at 950°C on matrix without Al etching (d_{pore}=80 nm)	Cu_930@AAO/Al_80	Cu	930

TABLE 7.2 *(Continued)*

No.	Sample	Designation	Doping atom	T_{ev}, °C
19	ZnS, doped by Mn at 1250°C on matrix without Al etching (d_{pore}=80 nm)	Mn_1200@AAO/Al_80	Mn	1200
20	ZnS, doped by Mn at 1350°C on matrix without Al etching (d_{pore}=80 nm)	Mn_1230@AAO/Al_80	Mn	1230
21	ZnS, doped by Cu at 900°C on Al_2O_3 matrix with etched Al (d_{pore}=80 nm)	Cu_900@AAO_80	Cu	900
22	ZnS, doped by Cu at 950°C on Al_2O_3 matrix with etched Al (d_{pore}=80 nm)	Cu_930@AAO_80	Cu	930
23	ZnS, doped by Mn at 1250°C on Al_2O_3 matrix with etched Al (d_{pore}= 80 nm)	Mn_1200@AAO_80	Mn	1200
24	ZnS, doped by Mn at 1350°C on Al_2O_3 matrix with etched Al (d_{pore}= 80 nm)	Mn_1230@AAO_80	Mn	1230

KEYWORDS

- **doped ZnS deposition**
- **experimental facility**
- **samples nomenclature**
- **synthesis of porous Al_2O_3**
- **nanostructures of ZnS**

REFERENCE

1. Schneider, C. A.; Rasband, W. S.; Eliceiri, K. W. NIH Image to ImageJ: 25 Years of Image Analysis. *Nat. Methods* **2012,** *9* (7), 671–675.

CHAPTER 8

Structure and Chemical Composition of Electroluminescent Nanocomposites ZnS:(Cu,Mn)@AAO

ABSTRACT

This chapter includes the results of complex experimental investigations of the structure and chemical composition of electroluminescent nanocomposites ZnS:(Cu,Mn)@AAO. Chemical composition and chemical bond character of such nanocomposites are discussed. The results of X-ray photoelectron spectroscopy experiments of samples alloyed with Cu and Mn are presented. Morphology of nanostructures based on raster electron microscopy data, X-ray diffraction investigations, and local atomic structure based on extended X-ray absorption fine structure (EXAFS)-spectroscopy data are considered. In addition, the results of EXAFS investigations of samples alloyed with Cu and alloyed with Mn and results of X-ray absorption near-edge structure investigations are described.

8.1 CHEMICAL COMPOSITION AND CHEMICAL BOND CHARACTER

The chemical bond characters were investigated to reveal the chemical interaction between the impurity and sulfur ions, the formation of chemical bonds between them can result in both the formation of separate phase and preservation of the phase composition when zinc atoms are replaced by the impurity atoms. Also, the segregation of doped atoms is possible on the sample surface with the formation of clusters.

The investigations were carried out on electron spectrometer SPECS in the mode of energy analyzer constant transmission energy of 15 eV using

MgKα radiation (E_{ph} = 1253.6 eV). The X-ray photoelectron spectroscopy (XPS) spectra of films deposited onto glass templates were obtained. It was assumed that the surface properties of films on glass surface and porous aluminum oxide surface do not differ much because of the formation of solid film after the material fills the matrix pores.

The spectra were taken both directly from the initial surface and after etching in spectrometer chamber with Ar^+ ions with the energy of 4 keV and ion current density of 30 mcA/cm^2 within 1 and 5 min. The etching rate was 1–1.2 nm/min. The chemical state of the elements was identified using reference data.[1] XPS experimental data were processed with the help of software package CasaXPS.[2]

The spectra were decomposed into components by the fitting using the mixture of Gaussian and Lorentz functions.

8.1.1 XPS of Cu-doped Samples

Overview XPS spectra of Cu_900_SiO_2 (a) and Cu_950_SiO_2 (b) samples before and after etching are demonstrated in Figure 8.1.

It is seen in the overview spectra that all atoms, whose presence was set up by initial deposition conditions are present in the films. The presence of oxygen and carbon peaks is connected with their adsorption from the atmosphere when transferring the samples from the deposition chamber to the spectrometer chamber.

Let us consider zinc, sulfur, and copper spectra in more detail (Figs. 8.2–8.5).

The positions of Zn and Cu maxima (Figs. 8.2 and 8.4) correspond to the availability of their bonds with sulfur, and intensity of maximums of zinc spectra correlates with copper content in the films; since the surface appears to be saturated with copper, zinc concentration is less.

The intensity of copper spectra decreases after etching, but it increases in zinc spectra. Most probably, it is connected with the fact that copper atoms segregate to the surface, but cuprous copper Cu^+ forms chemical bond with sulfur with the formation of Cu_2S in the whole film volume. The joint analysis of copper Auger and 2p lines (Fig. 8.1) demonstrates its absence in the pure form Cu^0.

Chemical state of the elements was evaluated by the spectra decomposition into components of the mixture of Gaussian and Lorentz functions.

The concentrations of chemical elements were calculated by the formula (4.12) using the areas under the corresponding peak.

Components of $Zn2p_{3/2}$, S2p, and Cu2p spectra for Cu_930_SiO_2 sample after etching within 5 min is given in Figure 8.6.

The concentration values of Zn, S, and Cu for Cu_900_SiO_2 and Cu_930_SiO_2 samples before and after etching are given in Table 8.1.

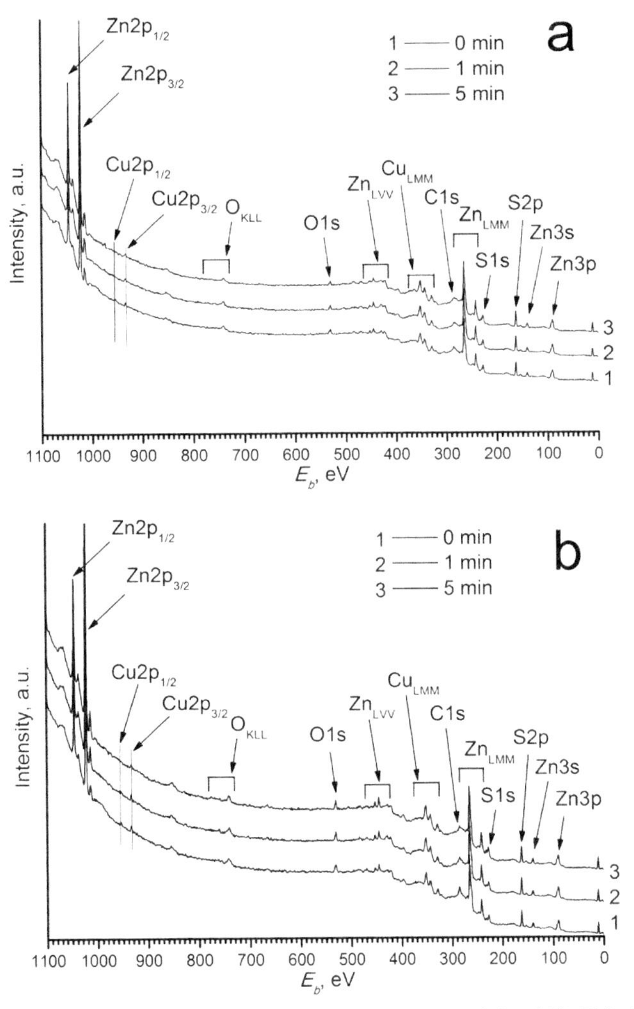

FIGURE 8.1 Overview XPS spectra of Cu_900_SiO_2 (a) and Cu_950_SiO_2 (b) samples before and after etching with argon ions.

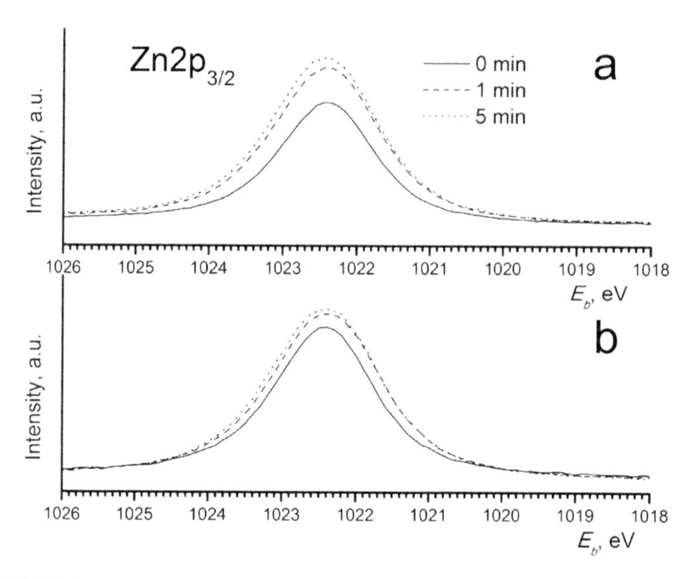

FIGURE 8.2 $2p_{3/2}$ spectra of Zn of Cu_900_SiO_2 (a) and Cu_930_SiO_2 (b) samples—initial and after etching with Ar^+ ions within 1 and 5 min.

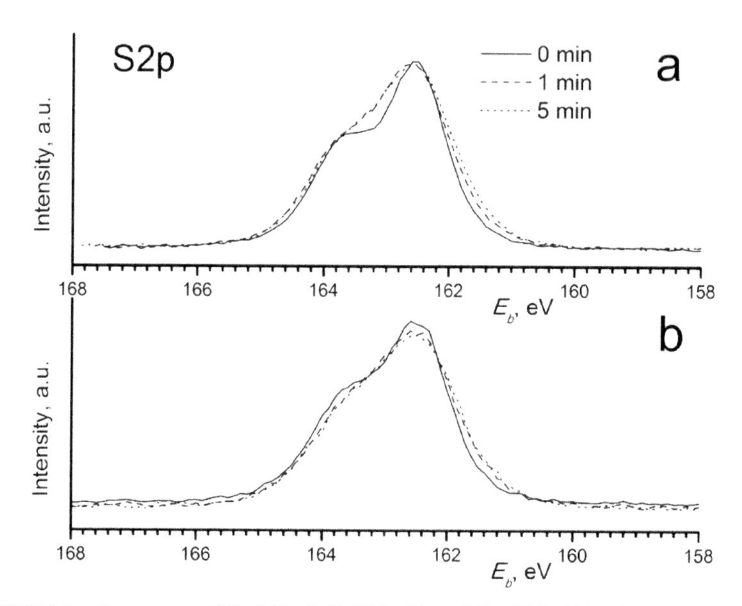

FIGURE 8.3 2p spectra of S of Cu_900_SiO_2 (a) and Cu_930_SiO_2 (b) samples—initial and after etching with Ar^+ ions within 1 and 5 min.

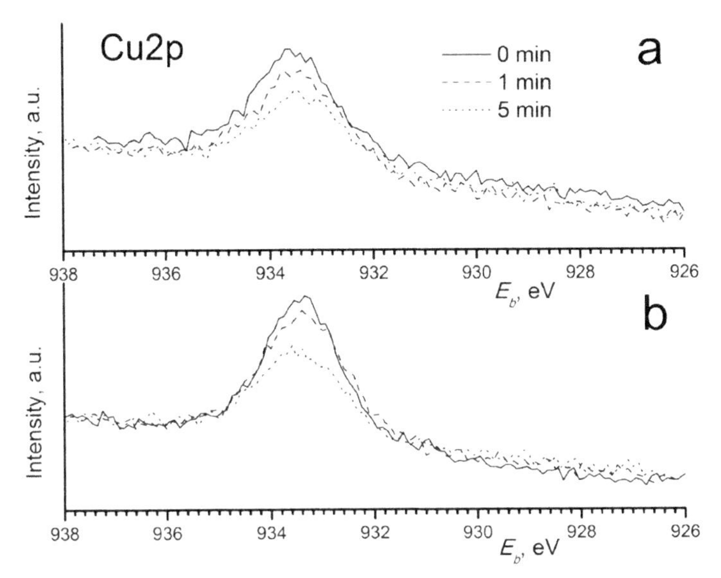

FIGURE 8.4 2p spectra of Cu of Cu_900_SiO_2 (a) and Cu_930_SiO_2 (b) samples—initial and after etching with Ar^+ ions within 1 and 5 min.

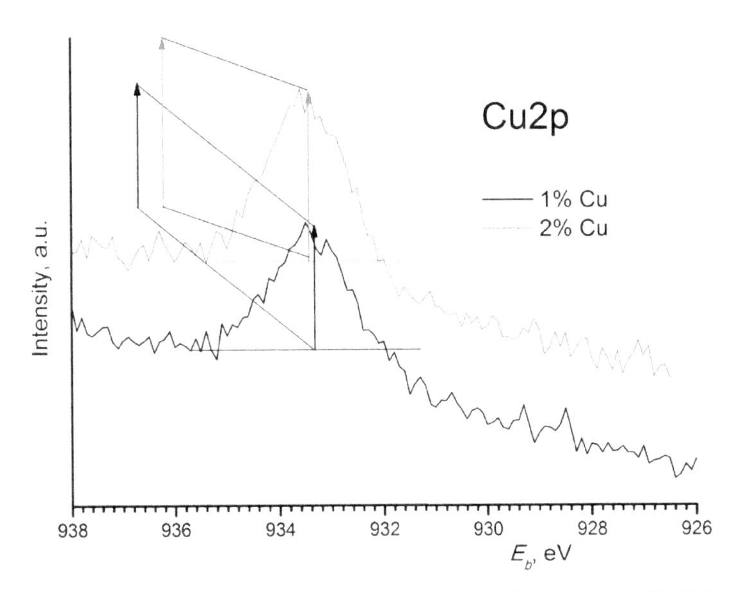

FIGURE 8.5 Comparison of intensities of 2p spectra of Cu of Cu_900_SiO_2 (black curve) and Cu_930_SiO_2 (red curve) samples with different percentage of Cu after etching with Ar^+ ions within 5 min.

TABLE 8.1 Zn, S, and Cu Concentrations in Cu_900_SiO_2 and Cu_930_SiO_2 Samples.

Sample	C_{Cu}, at.%	C_S, at.%	C_{Zn}, at.%
Cu_900_SiO_2 0 min	1,6(5)	50,2(5)	48,2(5)
Cu_900_SiO_2 1 min	0,8(5)	50,3(5)	48,9(5)
Cu_900_SiO_2 5 min	0,7(5)	63,7(5)	35,6(5)
Cu_930_SiO_2 0 min	2,4(5)	47,6(5)	50,0(5)
Cu_930_SiO_2 1 min	2,4(5)	47,0(5)	50,6(5)
Cu_930_SiO_2 5 min	1,4(5)	47,0(5)	51,6(5)

It is seen from Table 8.1 that copper concentration decreases with the etching time, that is, natural, since, as it was already pointed out before, copper actively segregates to the film surface. The concentration behavior of Cu_900_SiO_2 sample is also natural, where copper concentration on the surface after etching within 1 min correlates with zinc concentration. The data of rewashing with sulfur after 5 min of etching are, more likely, connected with the experimental statistic error.

The re-enrichment with zinc atoms is observed for Cu_930_SiO_2 sample; here, the influence of film deposition, which can differ from the case of Cu_900 samples deposition, since the ideal temperature control is impossible from experiment to experiment, and the template temperature insufficient increase (not more than 10°C) results in sulfur re-evaporation from the template back to the deposition chamber.

It should be pointed out that other investigation method, such as X-ray diffraction and extended X-ray absorption fine structure (EXAFS)-spectroscopy, sufficient structural changes from copper amount, were not revealed.

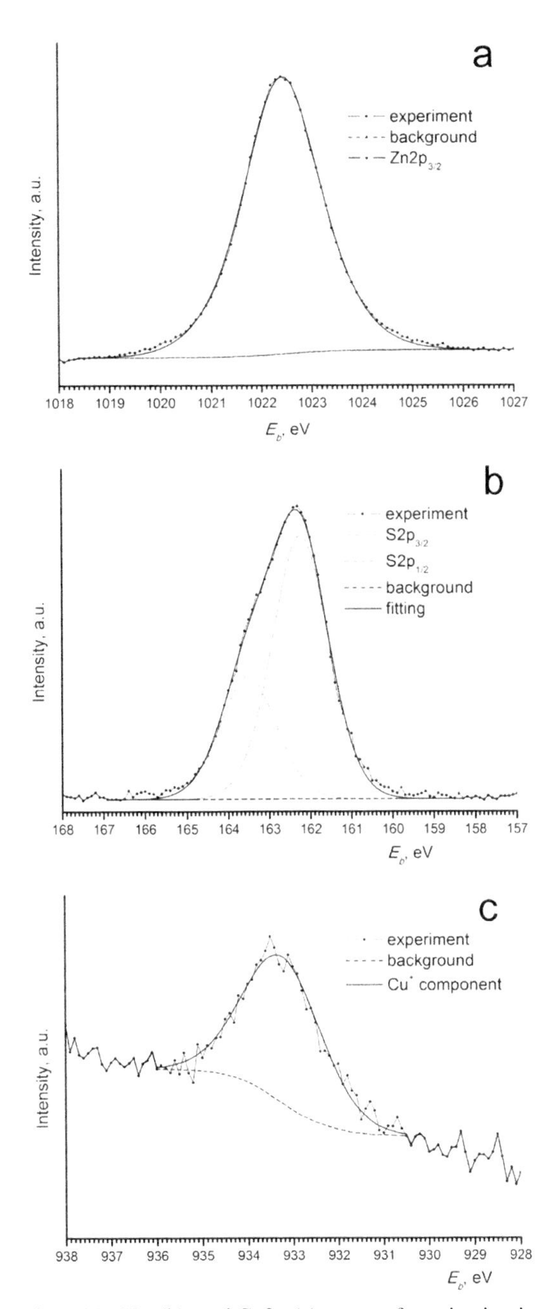

FIGURE 8.6 $Zn2p_{3/2}$ (a), S2p (b), and Cu2p (c) spectra factorization into components for Cu_930_SiO_2 sample after etching within 5 min.

8.1.2 XPS of Mn-doped Samples

Overview XPS spectra of Mn_1230_SiO_2 (a) and Mn_1200_SiO_2 (b) samples before and after etching are given in Figure 8.7.

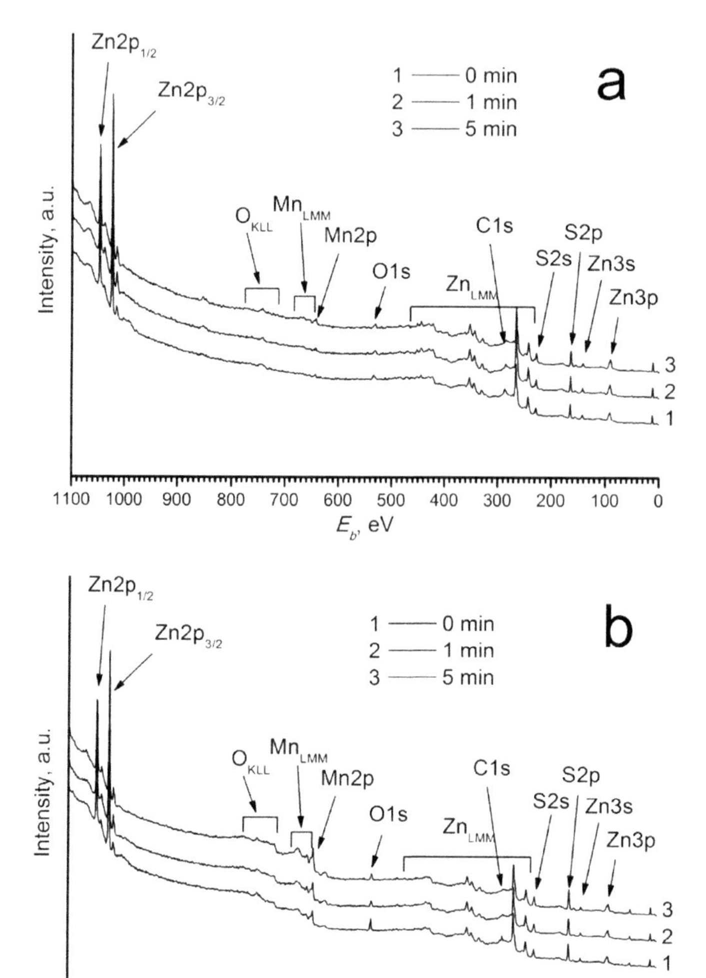

FIGURE 8.7 Overview XPS spectra of Mn_1230_SiO_2 (a) and Mn_1200_SiO_2 (b) samples before and after etching with argon ions.

In the overview spectra, it is seen that all the atoms, whose availability was set up by initial deposition conditions, are present in the films. The presence of carbon and oxygen peaks, as in the samples alloyed with copper, is connected with their adsorption from the atmosphere when transferring the samples from the deposition chamber to the spectrometer chamber.

Let us consider zinc, sulfur, and manganese spectra in more detail (Figs. 8.8–8.11).

The positions of Zn and Mn maxima correspond to the presence of their bonds with sulfur. All films are over-rich with sulfur since manganese alloying was carried out by joint deposition of manganese sulfide. At the same time, the intensity of zinc spectra maxima correlates with manganese content in the films. The intensity of manganese spectra increases after etching. Most probably, it is connected with the fact that during deposition of the evaporator gate, from which the compound with alloying element was deposited, was closed first. The "shake up" with greater bond energy to the right of the main 2p1/2 and 2p3/2 lines proves the bivalency of manganese ions (Mn^{2+}), that is, MnS compound is formed after the deposition, and Mn in the pure form is absent.

The concentrations of Zn, S, and Mn for Mn_1200_SiO_2 and Mn_1230_SiO_2 samples before and after etching given in Table 8.2 were calculated based on the areas under the peaks S2p, Mn2p, and Zn3d (Figs. 8.8–8.10). Mn concentration increase by the sample depth confirms the above assumption on earlier closing of the evaporator gate with MnS.

TABLE 8.2 Concentration of Zn, S, and Mn in Mn_1200_SiO_2 and Mn_1230_SiO_2 Samples.

Sample	C_{Mn}, at.%	C_S, at.%	C_{Zn}, at.%
Mn_1230_SiO_2 0 min	10,8	54,7	34,5
Mn_1230_SiO_2 1 min	12,6	55	32,4
Mn_1230_SiO_2 5 min	16,3	55,5	28,2
Mn_1200_SiO_2 0 min	1,6	52,3	46,2
Mn_1200_SiO_2 1 min	2,7	51,9	45,4
Mn_1200_SiO_2 5 min	5,9	52,3	41,8

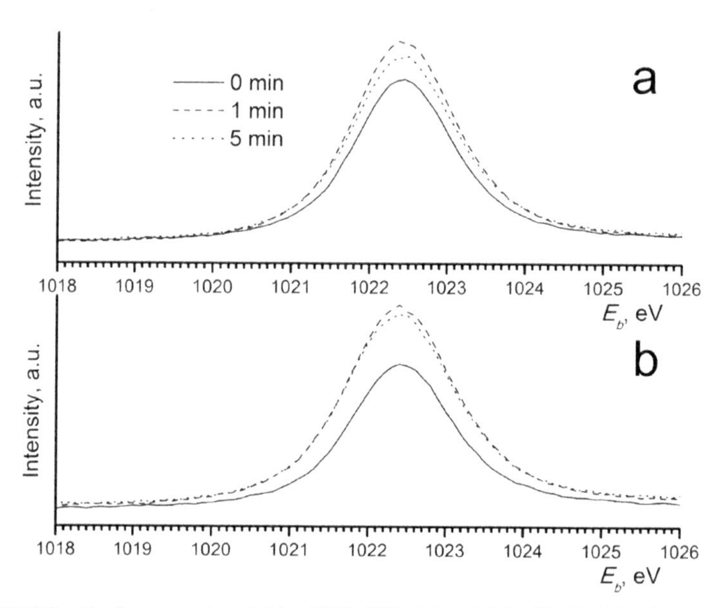

FIGURE 8.8 Zn $2p_{3/2}$ spectra of Mn_1230_SiO_2 (a) and Mn_1230_SiO_2 (b) samples—initial and after 1 and 5 min etching with Ar^+ ions.

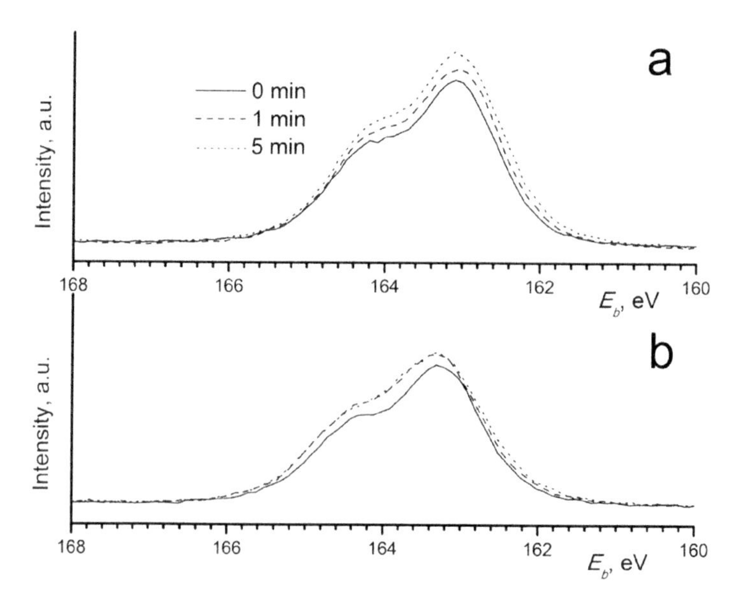

FIGURE 8.9 S 2p spectra of Mn_1230_SiO_2 (a) and Mn_1200_SiO_2 (b) samples—initial and after 1 and 5 min etching with Ar^+ ions.

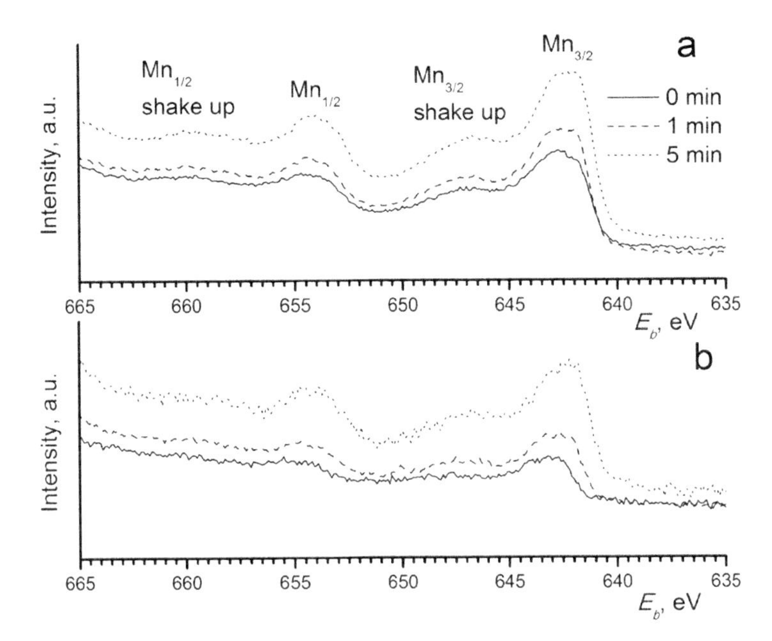

FIGURE 8.10 Mn 2p spectra of Mn_1230_SiO_2 (a) and Mn_1200_SiO_2 (b) samples—initial and after 1 and 5 min etching with Ar^+ ions.

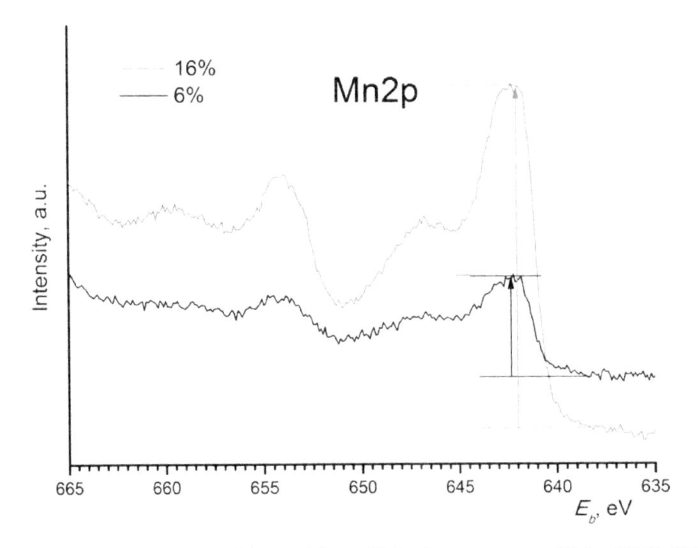

FIGURE 8.11 Comparison of intensities of Mn $2p_{3/2}$ spectra of Mn_1230_SiO_2 (black curve) and Mn_1230_SiO_2 (red curve) samples with different percentage of Mn after 5 min etching with Ar^+ ions.

8.2 MORPHOLOGY OF NANOSTRUCTURES BASED ON SCANNING ELECTRON MICROSCOPY DATA

Structure of cross-section cut of the samples and their surface morphology were investigated by scanning electron microscopy (SEM) by a Supra 50 VP instrument (LEO) equipped with energy-dispersive X-ray analysis system Oxford INCA Energy+.

The images of samples alloyed with copper with the concentration of 1 and 2 at.% and samples alloyed with manganese with concentration of 6 and 12 at.% are given in Figures 8.12 and 8.13 and Figures 8.14 and 8.15, respectively. The inserts illustrate the surface morphology, which has quite a significant roughness. Our previous investigations showed that ZnS penetrates into the pores up to 1 μm deep, but does not fill them forming a film on the matrix surface.[4] Under the applied deposition conditions, the thickness of these films is 600 nm, which is confirmed by the optical spectroscopy data. It is seen that the film is mainly formed on the pore walls; at the same time, it has a feebly-marked porous structure on the matrixes with the pore diameter of 40 nm. Porous films of ZnS are formed on the samples with pore diameter of 80 nm; moreover, the diameter of these pores practically equals the matrix pore diameter. This is especially seen in the sample Mn_1200@AAO_80. Unfortunately, these images confirm that we still cannot select the deposition modes and porous structure parameters of matrixes, which will result in filling the pores without the material film formation on the matrix surface.

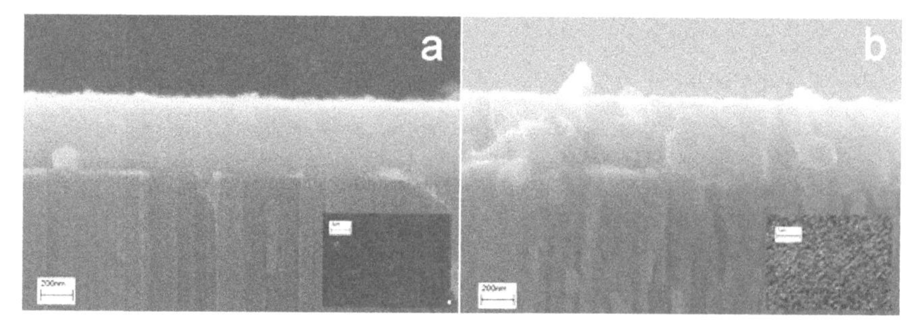

FIGURE 8.12 SEM images of cross-section cuts and surface of (a) Cu_900_@AAO_40 and (b) Cu_900_@AAO_80 samples.

FIGURE 8.13 SEM images of cross-section cuts and surface of Cu_930_@AAO_40 (a) and Cu_930_@AAO_80 (b) samples.

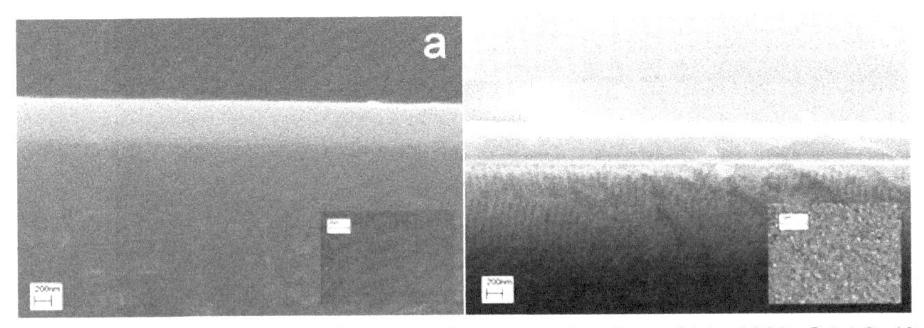

FIGURE 8.14 SEM images of cross-section cuts and surface of Mn_1200_@AAO_40 (a) and Mn_1200_@AAO_80 (b) samples.

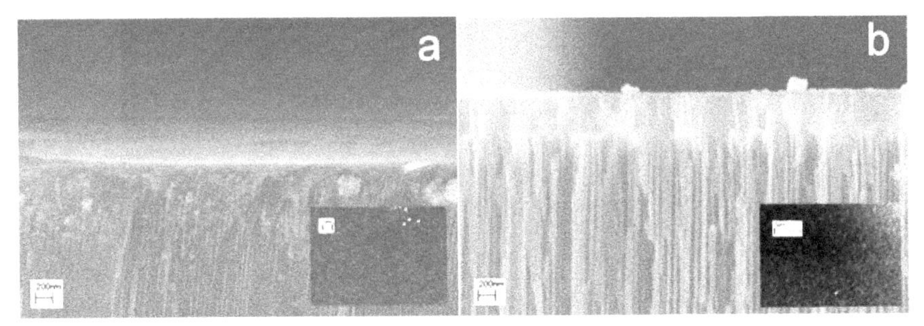

FIGURE 8.15 SEM images of cross-section cuts and surface of Mn_1230_@AAO_40 (a) and Mn_1230_@AAO_80 (b) samples.

8.3 X-RAY DIFFRACTION INVESTIGATIONS

The structural-phase state of the samples was investigated by the X-ray diffraction method on the diffractometer Rigaku MiniFlex 600 with Co-Kα excitation in the range of 2Θ angles from 10° up to 140° with the increment of 0.05°. The signal was accumulated in one point within 1.5 s.

The films obtained are mainly crystallized in the cubic phase (Fig. 8.16). However, the maximum characteristic for the reflection from the plane (100) of hexagonal syngony is observed to the left of the maximum corresponding to the reflection from the plane (111) of cubic syngony. Nevertheless, the most intensive in hexagonal syngony maximum (101) is missing in Cu_930@AAO/Al_40 (red curve) and Cu_930@AAO/Al_80 (blue curve) samples. Probably, the single hexagonal phase is not separated with a large copper amount. The maximum of hexagonal phase (101) for Cu_900@AAO/Al_40 sample (black curve) probably merges with the maximum (002) of Cu_2S, a weakly-intensive peak is observed for Cu_900@AAO/Al_80 sample (green curve).

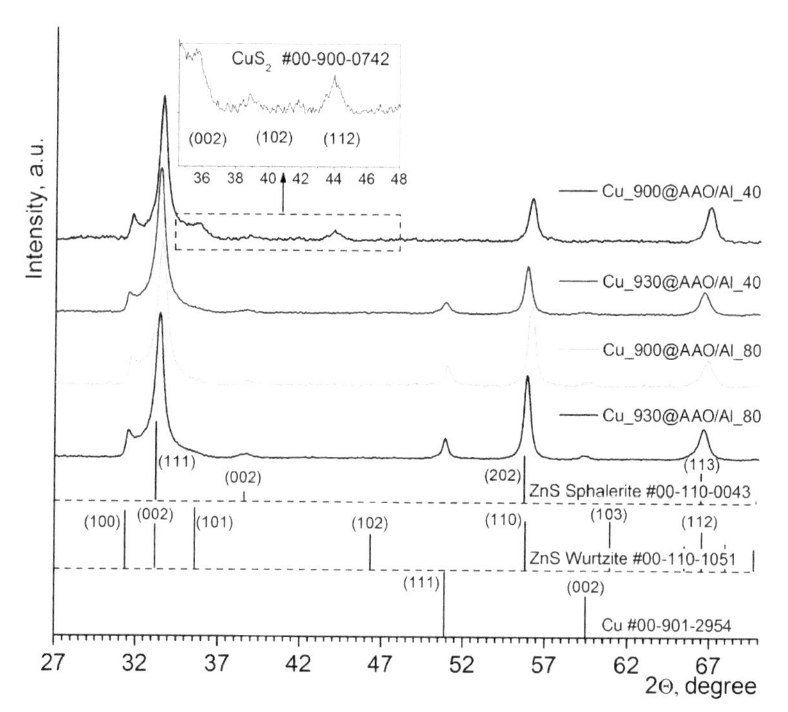

FIGURE 8.16 X-ray diffraction patterns of Cu-doped samples.

In the process of vacuum thermal deposition, the film, first of all, grows on defects. Because of porosity, the surface of anodic aluminum oxide has a regular defective structure, and nucleation of crystallites can be observed simultaneously on the whole surface. When the crystallites are twinning, the interlayer with disturbed order of packing can be formed in the process of further deposition. Since the structures of sphalerite and wurtzite are polytypes, it can be assumed that the interlayers formed have hexagonal structure, being the reason for the emergence of the reflex (100) of hexagonal syngony on diffraction pattern.

The metal copper phase is present on Cu_930@AAO/Al_40, Cu_930@AAO/Al_80, and Cu_900@AAO/Al_80 samples. The emergence of pure copper phase is conditioned by the energetically more favorable formation of ZnS in comparison with copper and sulfur composition. Thus, the standard formation enthalpy for ZnS is –205.4 kJ/mol, for CuS –53.14 kJ/mol, and for Cu_2S –79.5 kJ/mol. With this deposition method, the films obtained have excessive zinc content, and sulfur may not be enough for the binding with copper. It should be pointed out that based on the previous results of X-ray electron spectroscopy, copper can be completely bound with sulfur in the near-surface layers, and copper clusters can be formed at the "semiconductor/matrix" interface. Pure copper phase is missing on Cu_900@AAO/Al_40 sample, but the compound of copper and sulfur emerge. It is impossible to determine the stoichiometric composition of Cu_xS_y compound; however, CuS_2 (cubic lattice, spatial group Pa3) is the closest in reflex location from phase analysis. This sample differs in minimum pore sizes and low deposition rate by copper. Under the closed space conditions of the pores during the nanostructure formation, the segregation of copper atoms to the surface is limited and the probability of their interaction with sulfur increases. However, Cu_xS_y phase is missing on Cu_930@AAO/Al_40 sample obtained on the same matrix. The increase in copper deposition rate probably results in the formation of metal clusters with considerably lower chemical activity.

The change in the copper concentration in structures results in the change in interplanar distances, regardless of the formation of phase of pure copper or Cu_xS_y (Fig. 8.17). The interplanar distances increase with the copper concentration elevation. Such behavior can be connected with the penetration of copper atoms into interstitial cavities. If zinc atoms are replaced by copper atoms, the interplanar distances would decrease, since copper covalence radius is less than zinc covalence radius.

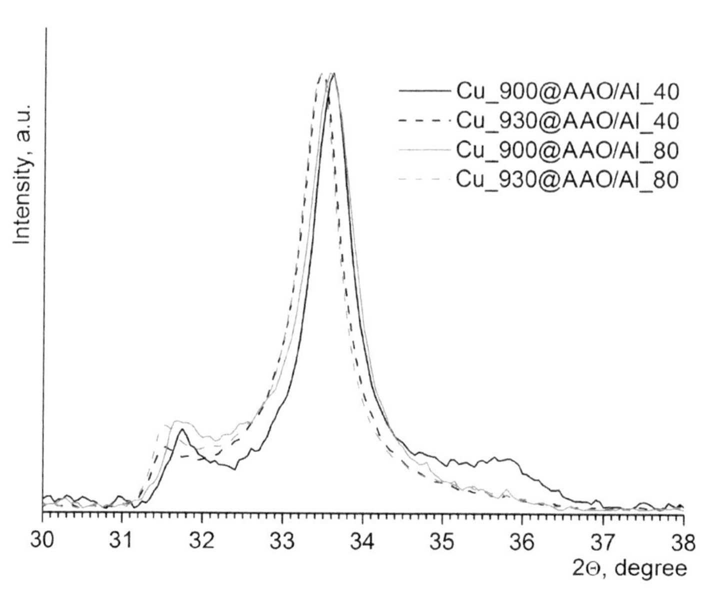

FIGURE 8.17 The magnified image of X-ray diffraction pattern in the region of maximum (111) of cubic phase of the Cu-doped samples.

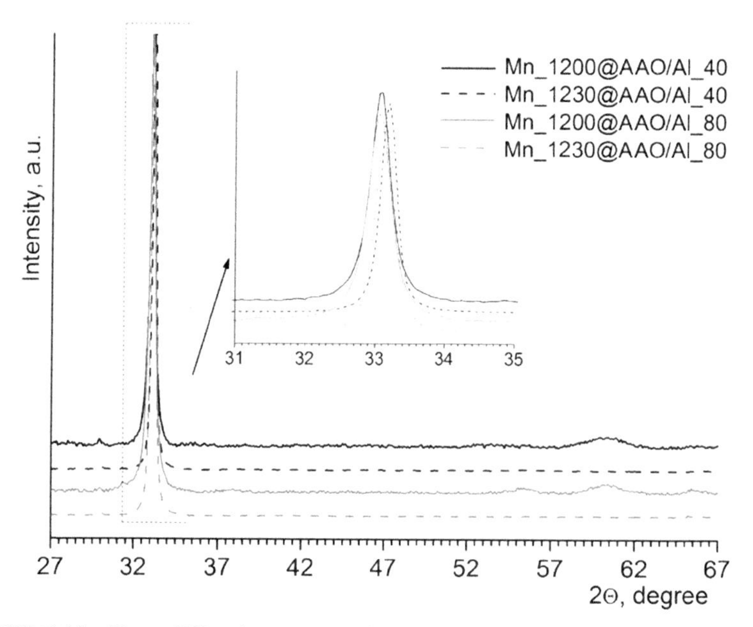

FIGURE 8.18 X-ray diffraction patterns of the Mn-doped samples.

During the deposition with manganese (Fig. 8.18), ZnS is crystallized only in the cubic phase with prominent texture in the plane (111). When the manganese concentration increases, the interplanar distances decrease. Such behavior can be connected with zinc replacement by manganese in the nodes of cubic lattice of ZnS. Such result does not contradict the data of X-ray electron spectroscopy, which indicate that manganese is bound with sulfur. At the same time, it is possible that excessive sulfur is concentrated in the interstices of ZnS lattice, which can be demonstrated by the method of EXAFS-spectroscopy.

8.4 LOCAL ATOMIC STRUCTURE BASED ON EXAFS-SPECTROSCOPY DATA

The absorption spectra on Zn K-edge (E_K=9659 eV for metal) were measured at the Structural Materials Science End-Station of Kurchatov synchrotron radiation source.[3] The electron beam energy of storage ring was 2.35 GeV and current of 80–100 mA. The X-ray absorption spectrum was registered on fluorescence mode, which was registered by the avalanche photodiode manufactured by FMB OXFORD.

The absorption spectra were processed with the help of software package Ifeffit1.2.11. After the standard procedures of background emphasizing, normalizing, and atom absorption separation μ_0, the Fourier transform of EXAFS—function μ in the interval of values of photoelectron wave vector of 3–12 $Å^{-1}$ for spectra measured on Zn K-edge with the weight function value of k^3 was performed. The crystalline structure of sphalerite was used as the initial model. The amplitude factor S_0^2 was taken equaled to 0.96 in all calculations. The coordination numbers equaled to 4, 12, and 12 for the first, second, and third coordination spheres, respectively, were also fixed.

8.4.1 Results of EXAFS Investigations of Cu-doped Samples

In Figures 8.19–8.21, the experimental Zn-K EXAFS-spectra, the normalized oscillating parts extracted from them, and their Fourier transforms for the copper-doped samples are presented.

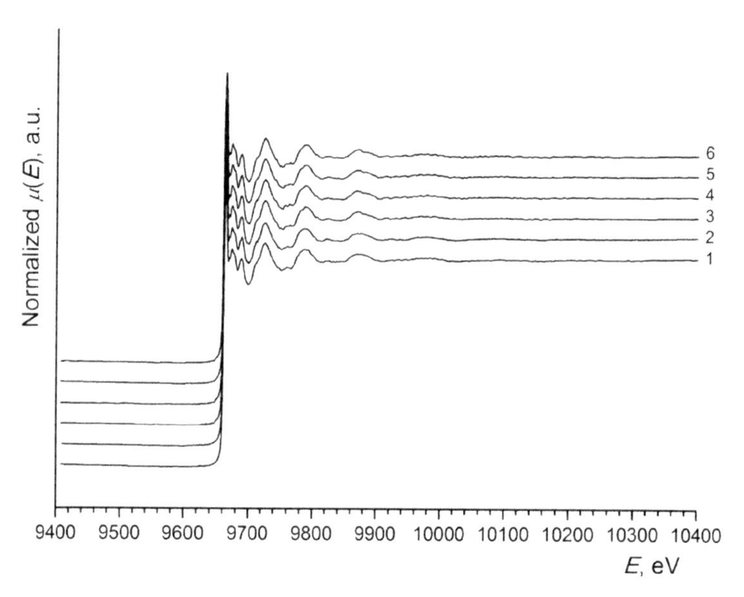

FIGURE 8.19 Experimental Zn-K EXAFS-spectra of the copper-doped samples: 1—Cu_900_SiO_2, 2—Cu_930_SiO_2, 3—Cu_900@AAO_40, 4—Cu_930@AAO_40, and 5—Cu_900@AAO_80, 1—Cu_930@AAO_80.

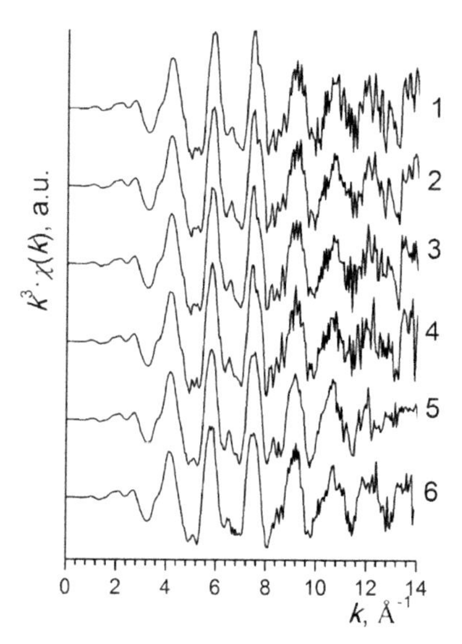

FIGURE 8.20 Normalized oscillating parts of EXAFS-spectra obtained on Zn absorption edge: 1—Cu_900_SiO_2, 2—Cu_930_SiO_2, 3—Cu_900@AAO_40, 4—Cu_930@AAO_40, and 5—Cu_900@AAO_80, 1—Cu_930@AAO_80.

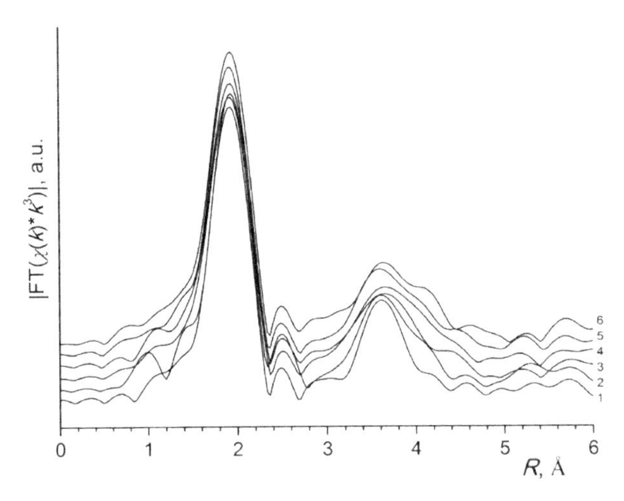

FIGURE 8.21 Fourier transforms of the normalized oscillating parts of EXAFS-spectra obtained on Zn absorption edge: 1—Cu_900_SiO_2, 2—Cu_930_SiO_2, 3—Cu_900@AAO_40, 4—Cu_930@AAO_40, and 5—Cu_900@AAO_80, 1—Cu_930@AAO_80.

It is very difficult to make certain conclusions by the view of absorption spectra and their normalized oscillating parts, whereas Fourier transforms have differences for different samples. The analysis of the peaks corresponding to sulfur distribution around zinc atoms in the first coordination sphere demonstrates the increased sulfur content in the samples obtained by the deposition onto glass (Cu_900_SiO_2 and Cu_930_SiO_2). For the sample with 1% copper, it does not contradict the data of X-ray electron spectroscopy, in the sample with 2 % Cu, sulfur concentration is less in accordance with XPS data. But in Cu_930_SiO_2 sample, the amplitude of the third peak, which also corresponds to sulfur distribution around zinc atoms but in the third coordination sphere, is lower and the number of sulfur atoms is nearly three times greater than in the first one. So, taking into account the influence of the long-range order, XPS data for Cu_930_SiO_2 sample are confirmed.

As for the samples obtained by the deposition onto porous aluminum oxide, the amplitudes of their Fourier transforms are lower than of the films on the glass but the contribution of the third coordination sphere is more vividly seen. Interatomic distances in all coordination spheres within errors (±0.01 Å—for the first and ±0.05—for the second and third) (Table 8.3) correspond to crystallographic ones for all samples but there is tendency to the radius increase of the second coordination sphere in

Cu_930@AAO_40 and Cu_900@AAO_40 samples in comparison with Cu_930@AAO_80 and Cu_930@AAO_80 samples.

TABLE 8.3 Interatomic Distances and Mean-Square Deviations of the Atoms of Zn Local Atomic Surroundings in 1–3 Coordination Spheres for the Cu-doped Samples. 1—Zn-S, 2—Zn-Zn, 3—Zn-S.

Sample	R_1, Å	σ_1^2, Å^2	R_2, Å	σ_2^2, Å^2	R_3, Å	σ_3^2, Å^2
ZnS crystal	2.34		3,82		4,48	
Cu_930@AAO_40	2.34	0.006	3.84	0.017	4.45	0.020
Cu_900@AAO_40	2.34	0.006	3.85	0.015	4.46	0.019
Cu_900@AAO_80	2.33	0.006	3.82	0.017	4.44	0.020
Cu_930@AAO_80	2.33	0.006	3.83	0.018	4.44	0.019
Cu_900_SiO_2	2.34	0.006	3.83	0.017	4.43	0.030
Cu_930_SiO_2	2.34	0.006	3.84	0.016	4.43	0.020

8.4.2 Results of EXAFS Investigations of Mn-doped Samples

In Figures 8.22–8.24, you can see the experimental Zn-K EXAFS-spectra, the normalized oscillating parts extracted from them, and their Fourier transforms for the samples alloyed with manganese.

The same as for the copper-doped samples, it is difficult to make certain conclusions by the view of absorption spectra and their normalized oscillating parts, whereas Fourier transforms have differences for different samples. The amplitude of the peak corresponding to the first coordination sphere (ZnS) depends on manganese concentration; for the samples obtained at MnS evaporation temperature of 1200°C, it is much lower regardless of the template type.

This is natural as the concentration of additional sulfur emerging because of MnS deposition is less than for the samples obtained at MnS evaporation temperature of 1230°C. This result correlates well with XPS data. But the situation is not that definite depending on the pore diameter. Thus, the peak amplitude of the first coordination sphere of Mn_1200@AAO_80 sample is higher than of Mn_1200@AAO_40 sample. The situation for Mn_1230@AAO_80 and Mn_1230@AAO_40 samples is reverse.

This is possibly connected with the samples morphology; on the basis of the results of SEM investigations, the film formed on the surface of aluminum oxide matrix with the pore diameter of 80 nm is not solid but repeats the matrix surface morphology.

The interatomic distances, in contrast to the samples alloyed with copper, correspond to crystallographic values only in the first and second coordination spheres, whereas the interatomic distances in the third coordination sphere are less (Table 8.4).

This is connected with the fact that the size of manganese atom is less than that of zinc; when being embedded into the lattice, ZnS parameter of the lattice decreases, thus resulting in the interatomic distance reduction.

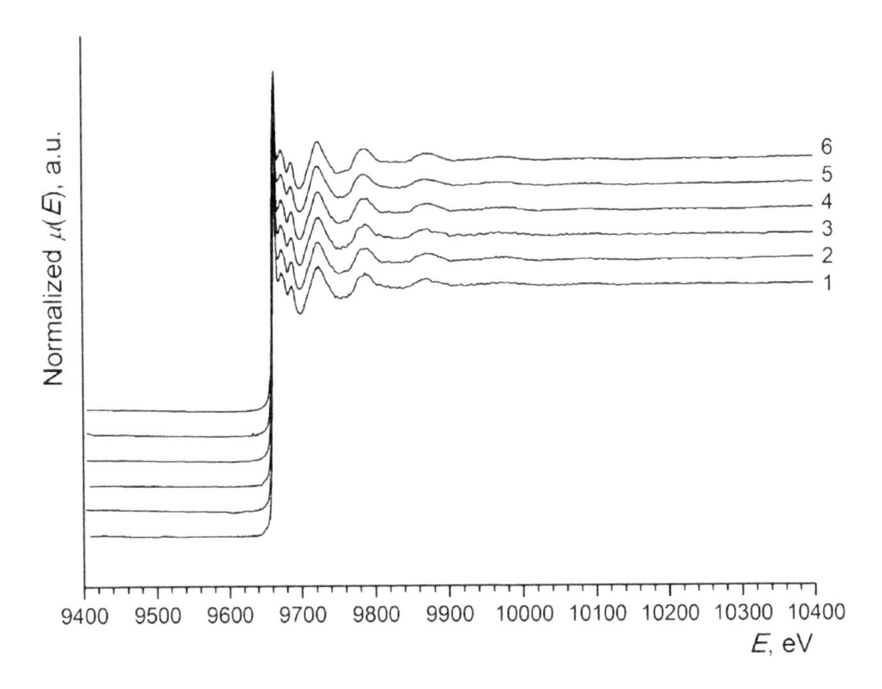

FIGURE 8.22 Experimental Zn-K EXAFS-spectra of the manganese-doped samples: 1—Mn_1200@AAO_40, 2—Mn_1230@AAO_40, 3—Mn_1200@AAO_80, 4—Mn_1230@AAO_80, 5—Mn_1200_SiO_2, and 6—Mn_1230_SiO_2.

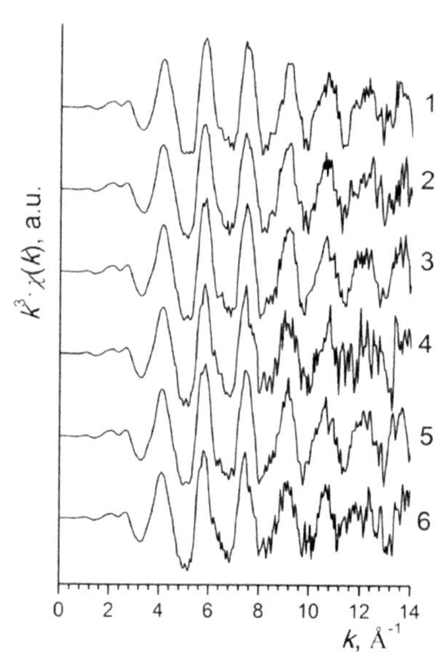

FIGURE 8.23 Normalized oscillating parts of EXAFS-spectra obtained on Zn absorption edge: 1—Mn_1200@AAO_40, 2—Mn_1230@AAO_40, 3—Mn_1200@AAO_80, 4—Mn_1230@AAO_80, 5—Mn_1200_SiO_2, and 6—Mn_1230_SiO_2.

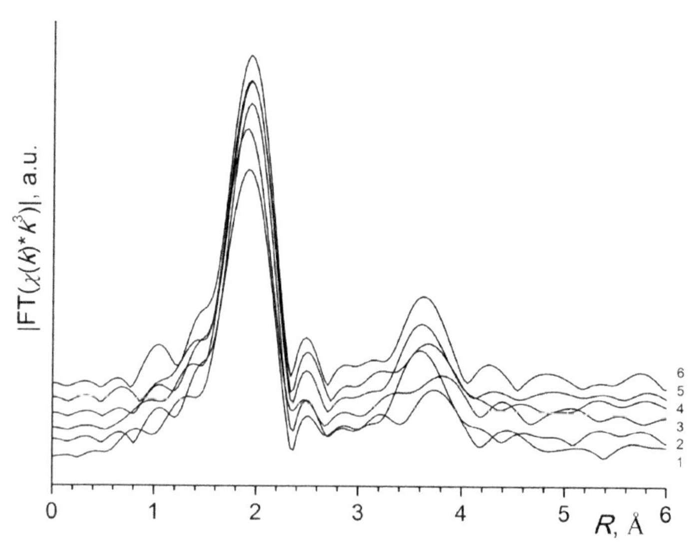

FIGURE 8.24 Fourier transforms of the normalized oscillating parts of EXAS-spectra obtained on Zn absorption edge: 1—Mn_1200@AAO_40, 2—Mn_1230@AAO_40, 3—Mn_1200@AAO_80, 4—Mn_1230@AAO_80, 5—Mn_1200_SiO_2, and 6—Mn_1230_SiO_2.

TABLE 8.4 Lengths of Chemical Bonds and Mean-square Deviations of the Atoms of Zn Local Atomic Surroundings in 1–3 Coordination Spheres for the Mn-doped Samples. 1—Zn-S, 2—Zn-Zn, and 3—Zn-S.

Sample	R_1, Å	σ_1^2, Å^2	R_2, Å	σ_2^2, Å^2	R_3, Å	σ_3^2, Å^2
ZnS crystal	2.34		3.82		4.48	
Mn_1230@AAO_40	2.34	0.006	3.84	0.020	4.40	0.045
Mn_1230@AAO_80	2.34	0.006	3.84	0.020	4.39	0.047
Mn_1200@AAO_40	2.34	0.007	3.83	0.024	4.46	0.052
Mn_1200@AAO_80	2.33	0.007	3.84	0.023	4.41	0.033
Mn_1200_SiO_2	2.34	0.006	3.84	0.025	4.29	0.051
Mn_1230_SiO_2	2.34	0.006	3.85	0.022	4.46	0.051

8.4.3 Results of XANES Investigations

The experimental X-ray absorption near-edge structure (XANES) spectra obtained are given in Figures 8.25 and 8.26, XANES spectra of all samples measured on Zn K-edge practically coincide. This means that there are no significant distortions of the crystalline lattice when the Mn and Cu atoms are imbedded. The slight changes in the intensity of "the white line" for the spectra of ZnS copper-doped films are connected with their less integral thickness in comparison with the films on the porous surface of aluminum oxide. Fine changes in the spectra of the Mn-doped samples are connected with the availability of unbound sulfur atoms in the interplanar space of the crystalline lattice, resulting in barely visible local distortions.

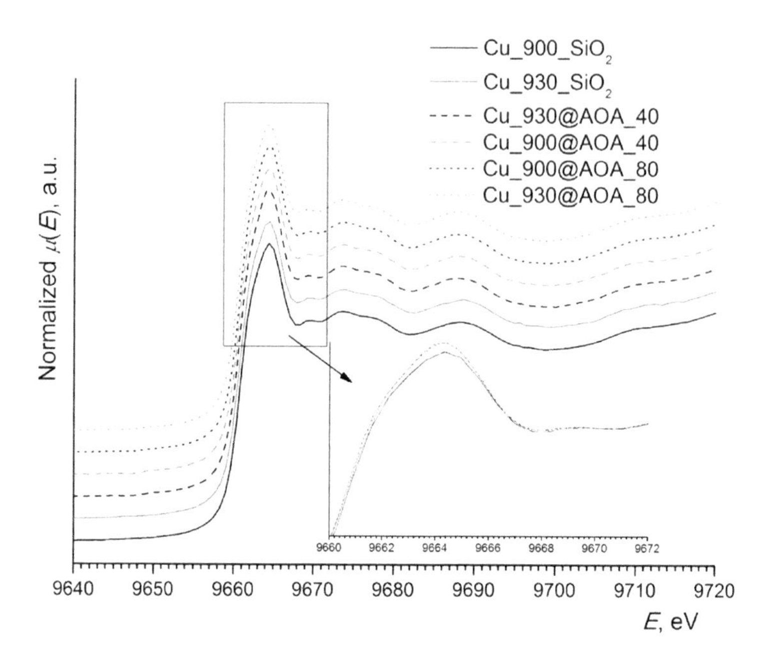

FIGURE 8.25 Zn-K XANES-spectra of the Cu-doped samples.

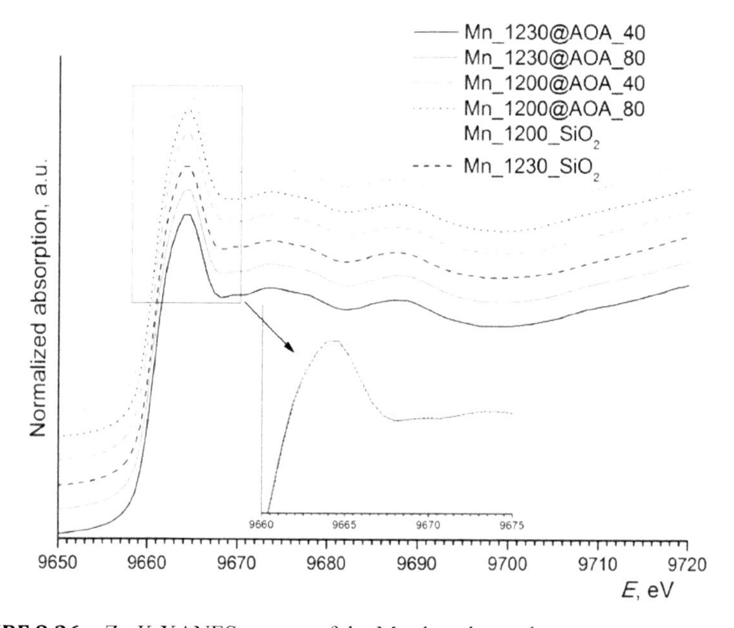

FIGURE 8.26 Zn-K XANES-spectra of the Mn-doped samples.

KEYWORDS

- **chemical bonds and composition**
- **extended X-ray absorption fine structure-spectroscopy**
- **scanning electron microscopy**
- **structure**
- **X-ray absorption near-edge structure-spectroscopy**
- **X-ray photoelectron spectroscopy study**
- **X-ray diffraction**

REFERENCES

1. Wagner, C. D.; Riggs, W. M.; Davis, L. E.; Moulder, J. F. *Handbook of X-Ray Photoelectron Spectroscopy*; Perkin-Elmer Corporation: USA, 1979; p 192.
2. Fairley, N. *CasaXPS Manual;* Casa Sofware: USA, 2010; p 176.
3. Chernyshov A. A.; Veligzhanin A. A.; Zubavichus Y. V. Structural Materials Science End-station at the Kurchatov Synchrotron Radiation Source: Recent Instrumentation Upgrades and Experimental Results. *Nucl. Instr. Meth. Phys. Res.* **2009,** *A* 603, 95–98.
4. Valeev, R.; Romanov, E.; Beltukov, A.; Mukhgalin, V.; Roslyakov, I.; Eliseev, A. Structure and Luminescence Characteristics of ZnS Nanodot Array in Porous Anodic Aluminum Oxide. *Phys. Stat. Sol. C* **2012,** *9*, 1462–1465.

CHAPTER 9

Optical Properties of Electroluminescent Nanostructures of Cu and Mn-Doped ZnS

ABSTRACT

In the chapter, the experimental results of optical properties of electroluminescent nanostructures of ZnS alloyed with Cu and Mn ions are presented. Absorption spectra and optical band gap and electroluminescence are studied.

9.1 ABSORPTION SPECTRA AND OPTICAL BAND GAP

Optical properties of the samples were investigated using the optical spectrometer based on monochromator on diffraction grids MDR-41 (Experimental Designing Bureau "Spektr," Saint-Petersburg, Russia). The spectrometer is equipped with mirror condenser ZK-125, light filter unit with modulator BSM-12, set of diffraction grids to work in the spectral range of 0.2–2.0 μm (four pieces), set of light filters (five pieces), receiving unit operating in the spectral range from 200 up to 750 nm, receiving unit operating in the spectral range from 400 up to 1200 nm, receiving unit operating in the spectral range from 1000 up to 2500 nm, light source based on halogen lamp with power supply unit. All equipment is mounted on the optical bench. It is controlled by the hardware and software system installed in the microprocessor unit. The optical scheme of spectrometer is given in Figure 5.3.

The optical transmission spectra were registered in the wavelength range from 300–700 nm, then they were converted into absorption spectra with the ratio of A = 1 – T, where A—absorption spectrum and T—transmission spectrum of the sample. Before obtaining the sample spectrum, the transmission spectra of the templates were registered, which were later taken into account (deducted) when processing the samples spectra.

Figures 9.1 and 9.2 show the optical absorption spectra of samples obtained by deposition on glass substrates and matrixes of porous aluminum oxide with removed aluminum. The oscillating structure connected with interference during re-reflection at the interface "film/substrate" is observed on all samples. The analysis of oscillations on transmission spectra of the samples obtained on glass by the method given in Figure 9.3 allowed finding the coating thickness. Here, k_i—virtual part of the film refracting index on the region from $\lambda_{i\ min}$ to $\lambda_{i\ max}$, d_i—optical thickness, n_i—refracting index at the wavelength $\lambda_{i\ max}$. Then, d_i values were averaged. The thickness values are given in Table 9.1.

TABLE 9.1 Values of Optical Thicknesses of the ZnS:Cu(Mn) Films Obtained by Deposition on Glass.

Sample	Optical thickness d, nm
Cu_900_SiO_2	647 ± 55
Cu_930_SiO_2	572 ± 53
Mn_1200_SiO_2	548 ± 57
Mn_1230_SiO_2	493 ± 56

The sample spectra obtained on the substrates with bigger pore diameter (the samples are marked as AAO_80) have significant differences from the spectra of other samples. The absorption maximum shifts to the long-wave region, the oscillations are less marked.

This is connected both with the film discontinuity (see the results of SEM investigations, e.g., Figs. 8.12b–8.15b) and the influence of matrix pore diameter.

The films of impurity-doped ZnS on the surface of samples marked with AAO_40 are continuous (Figs. 8.12a–8.15a); therefore, the interference during re-reflection at the interface "film/substrate" is more distinct.

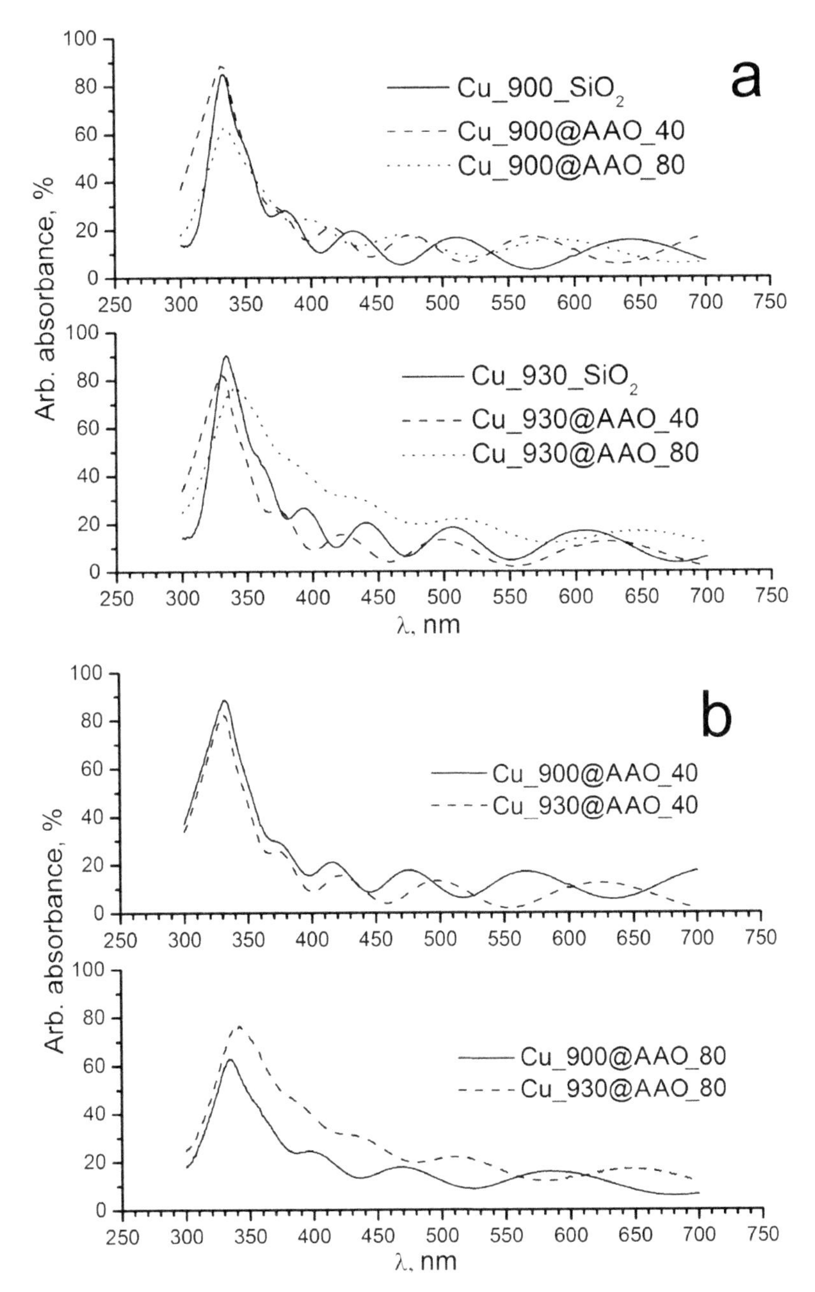

FIGURE 9.1 Absorption spectra of the samples of films and nanostructures of Cu-doped ZnS obtained at the same temperatures (a) and on the same porous substrates (b).

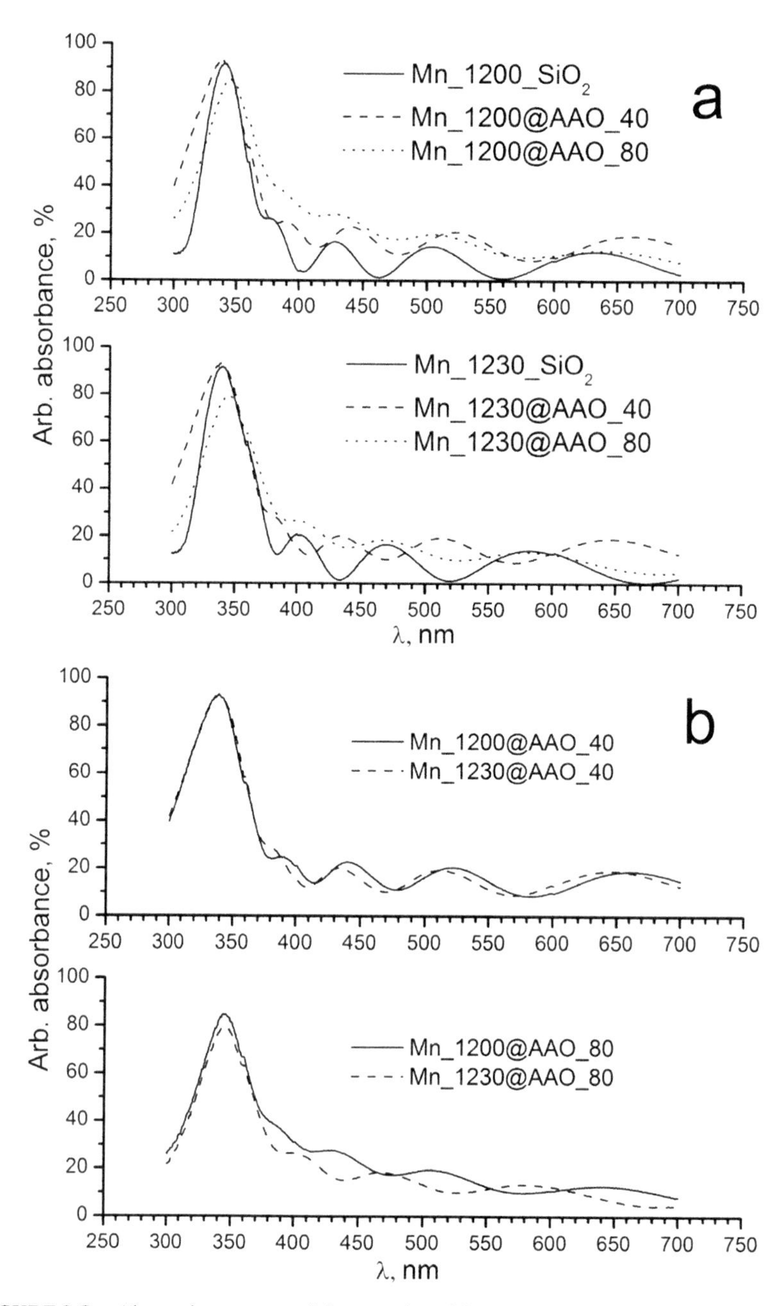

FIGURE 9.2 Absorption spectra of the samples of films and nanostructures of Mn-doped ZnS obtained at the same temperatures (a) and on the same porous substrates (b).

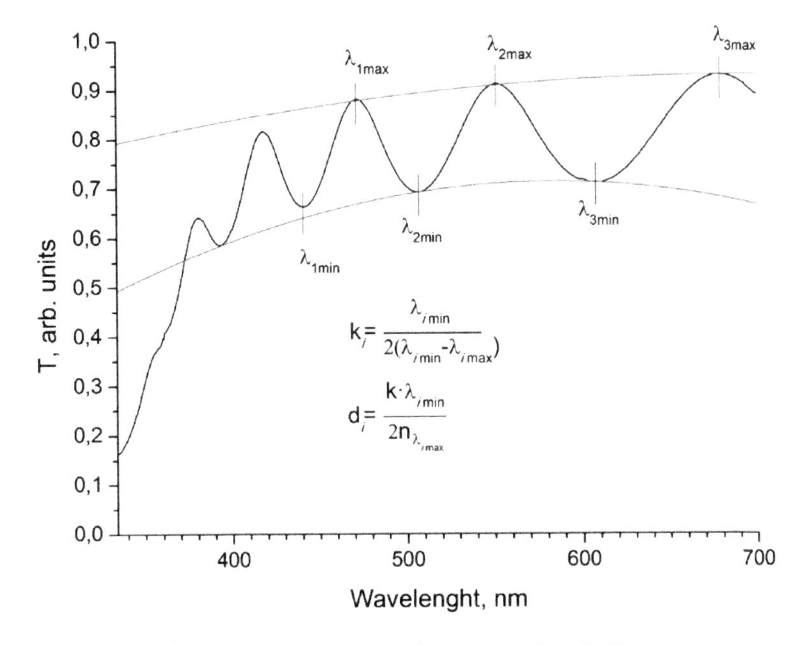

FIGURE 9.3 To the determination of optical thickness of the ZnS:Cu(Mn) films.

In comparison with films on glass, the maximums of sample fundamental absorption on porous aluminum oxide are broad; besides, the more is the diameter, the broader they are (Table 9.2).

TABLE 9.2 Values of Full Width at Half Maximums (FWHM) of Optical Absorption.

Cu-doped Sample	FWHM, nm	Mn-doped Sample	FWHM, nm
Cu_900_SiO_2	31	Mn_1200_SiO_2	33
Cu_930_SiO_2	31	Mn_1230_SiO_2	34
Cu_900@AAO_40	33	Mn_1200@AAO_40	38
Cu_930@AAO_40	39	Mn_1230@AAO_40	38
Cu_900@AAO_80	39	Mn_1200@AAO_80	42
Cu_930@AAO_80	50	Mn_1230@AAO_80	46

The optical band gap of the samples was evaluated following the technique described in paragraph 5.2 (Fig. 9.4). The values are given in Table 9.3.

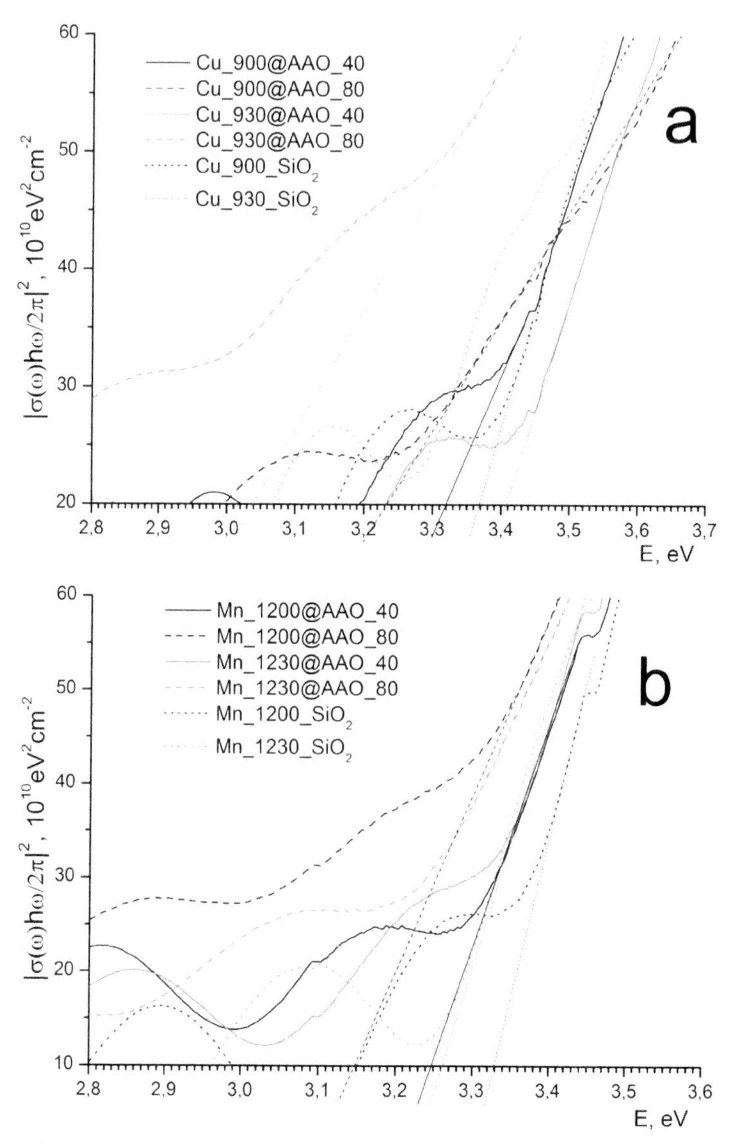

FIGURE 9.4 To the determination of optical band gap of the samples of films and Nanostructures of Cu-doped (a) and Mn-doped (b) ZnS.

TABLE 9.3 Values of Optical Band Gap of the Samples of Films and Nanostructures of Cu- and Mn-doped ZnS.

Cu-doped Sample	E_g, eV	Mn-doped Sample	E_g, eV
Cu_900_SiO_2	3.37	Mn_1200_SiO_2	3.33
Cu_930_SiO_2	3.28	Mn_1230_SiO_2	3.26
Cu_900@AAO_40	3.32	Mn_1200@AAO_40	3.25
Cu_930@AAO_40	3.11	Mn_1230@AAO_40	3.16
Cu_900@AAO_80	3.23	Mn_1200@AAO_80	3.24
Cu_930@AAO_80	3.04	Mn_1230@AAO_80	3.14

On the basis of the presented data, the dependence of E_g on the matrix pore diameter and sample composition is obvious. Thus, the band gap value decreases with the pore diameter increase in comparison with the films, the same way as with the increase in the ion concentration of the alloying element. The correlation with the full width at half maximums of absorption is also observed: the half-width increases with E_g decrease. This is possibly connected with the increased number of absorbing centers, which results in the valence and conductivity region blurring to decreasing the band gap.

9.2 ELECTROLUMINESCENCE

The electroluminescence spectra in the wavelength band from 250 up to 700 nm were registered on the optical spectrometer based on monochromator on diffraction grids MDR-41 (Experimental Designing Bureau "Spektr," Saint Petersburg, Russia). The radiation was excited by the electric field with the frequency of 50 Hz and voltage of 220 V.

ZnS:Cu(Mn) samples obtained by deposition on glass with transparent-conductive layer (Fig. 9.5 on the right) and on matrixes of porous aluminum oxide without removing aluminum base (Fig. 9.5 on the left) were taken for electroluminescence investigation. Then the buffer dielectric quartz layer and transparent-conductive ITO layer were deposited on the semiconductor surface. At the final stage, aluminum contact pads were applied. The samples were mounted on the stand (Fig. 5.8) and connected to AC power source (Figs. 5.9 and 5.10). The stand with the sample was fixed with the radiating surface as close as possible to the inlet slot of the monochromator MDR-41. The photo image of fluorescent sample is given in Figure 9.6.

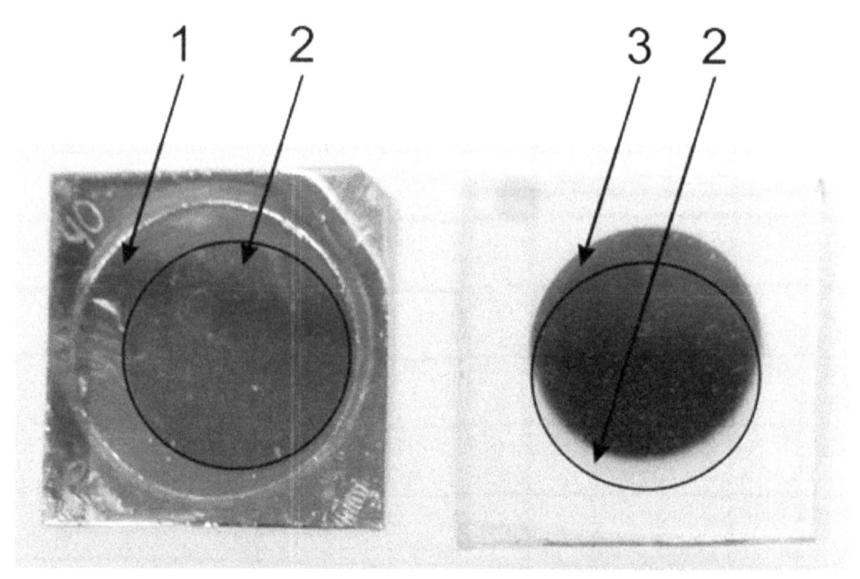

FIGURE 9.5 Photo images of typical samples for electroluminescence investigation. 1—AOA matrix, 2—luminophor, and 3— ITO film.

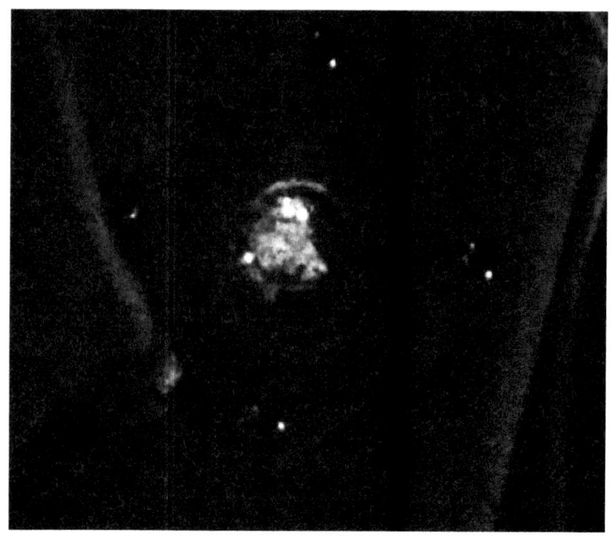

FIGURE 9.6 Photo image of Cu_930@AAO/Al_40 sample luminescence.

The electroluminescence spectra of the Cu-doped samples are given in Figure 9.7. Three luminescence bands with the maximums at 451 nm (x), 518 nm (y), and 542 nm (z) can be pointed out on them. According to the data we published before, the emission maxima with the wavelengths

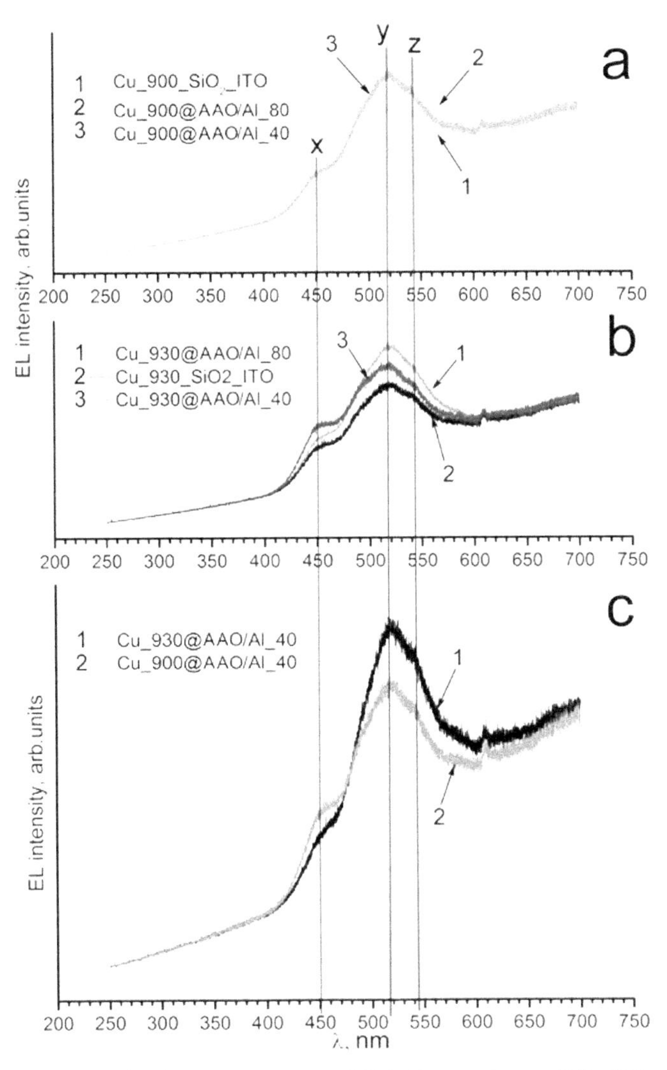

FIGURE 9.7 Electroluminescence spectra of ZnS:Cu samples on SiO_2_ITO and AAO/Al templates deposited at Cu evaporation temperatures of 900°C (a) and 930°C (b), as well as the spectra comparison for the samples obtained on AAO/Al_40 substrates at Cu evaporation temperatures of 900°C and 930°C (c).

of 451 and 518 nm are connected with the recombination of the charge carriers on sulfur and zinc vacancies, respectively,[1] but the maximum with the wavelength of 542 nm is connected, most likely, with the dissipation of the carriers from the conductivity region on the defective layers t_2 of copper.[2] It is seen that the emission intensity depends both on the phosphor composition and pore diameter of the matrix substrate: the admixture concentration and pore diameter increase results in the enhanced light emission intensity. This is probably connected with the increased intensity of electric field on nanostructures that results in more intensive light on them. There were no significant changes in the emission spectra from the material. On the basis of the data obtained, we can represent the structure of energy levels demonstrated in Figure 9.8.

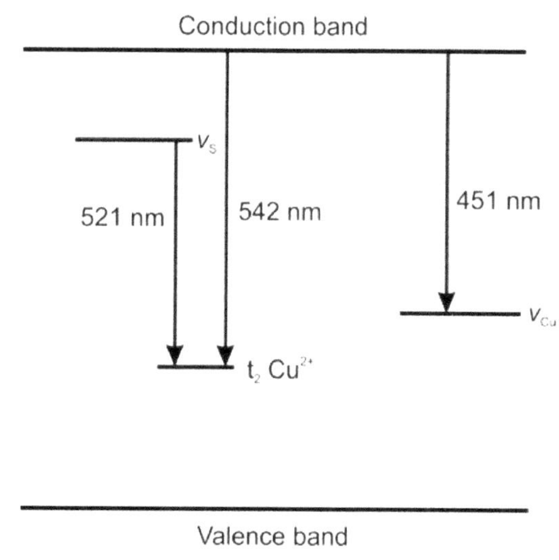

FIGURE 9.8 Structure of energy layers of ZnS:Cu phosphor.

Figure 9.9 shows the electroluminescence spectra of the Mn-doped samples. Two luminescence bands can be pointed out on them with the maximums at 530 nm (x) and 590 nm (y). According to the literature data, the emission with the wavelength of 530 nm corresponds to the self-activated luminescence of ZnS.[3,4]

The emission with the wavelength of 590 nm is connected with the emission on Mn^{2+} ions.[5] The structure of energy levels obtained based on the collected data is given in Figure 9.10.

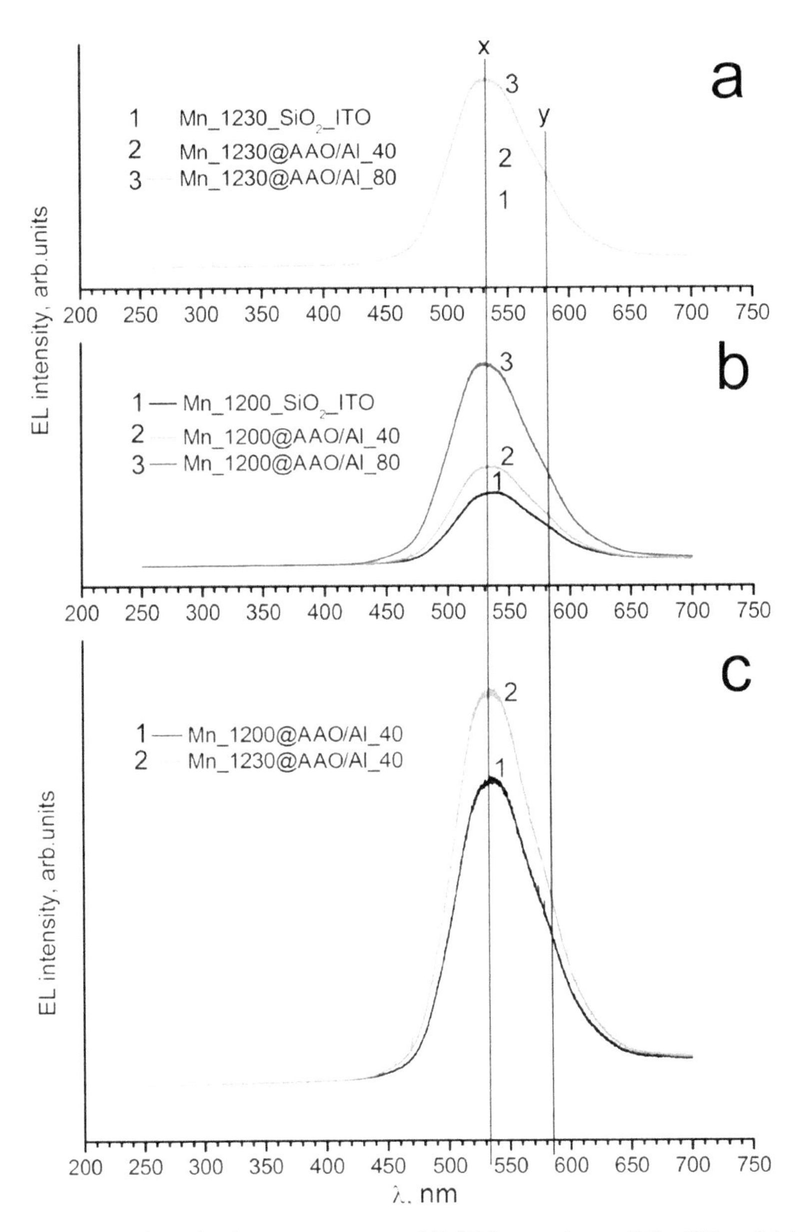

FIGURE 9.9 Electroluminescence spectra of ZnS:Mn samples on SiO_2_ITO and AAO/Al substrates deposited at Mn evaporation temperatures of 1200°C (a) and 1230°C (b), as well as the spectra comparison for the samples obtained on AAO/Al_40 at Mn evaporation temperatures of 1200°C and 1230°C (c).

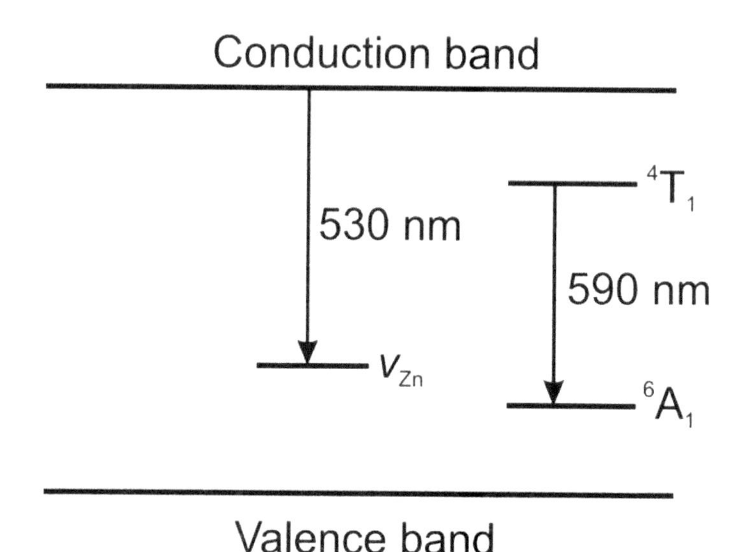

FIGURE 9.10 Structure of energy layers of ZnS:Mn phosphor.

KEYWORDS

- **ZnS:Cu (Mn) band gap**
- **ZnS:Cu(Mn) electroluminescence**
- **ZnS:Cu(Mn) optical absorption**
- **optical thickness**
- **absorption spectra**

REFERENCES

1. Valeev, R. G.; Petukhov, D. I.; Chukavin, A. I.; Beltiukov, A. N. Light-Emitting Nanocomposites on the Basis of ZnS:Cu Deposited into Porous Anodic Al_2O_3 Matrices. *Semiconductors* **2016,** *50* (2), 266–270.
2. Murugadoss, G. Synthesis and Photoluminescence Properties of Zinc Sulfide Nanoparticles Doped with Copper Using Effective Surfactants. *Particuology* **2013,** *11*, 566–573.
3. Voronov, Yu. V. Cathodo- and Photoluminescence of Zinc Sulfide in the UV Spectral Region, *Tr. Fiz. Inst. im. P. N. Lebedeva, Akad. Nauk. SSSR* **1973,** *68*, 3–94.

4. Morozova, N. K.; Kuznetsov, V. A.; Fock M. V. *Zinc Sulfide: Preparation and Optical Properties*; Nauka: Moscow, 1987; p 199 (In Russian).
5. Polezhaev, B. A.; Prokof'ev, T. A.; Kovalenko, A. V.; Bulanyi, M. F. Temperature Dependence of Photoluminescence and Electroluminescence of ZnS:Mn Crystals. *J. Appl. Spectrosc.* **2006,** *73* (5), 707–713.

CHAPTER 10

Results of Modeling the Deposition Processes of Nanofilms onto Aluminum Oxide Templates

ABSTRACT

The setting up of the problem of nanofilm deposition onto the templates of porous aluminum oxide is given in this chapter. The variants of epitaxial burying of porous templates based on aluminum oxide by atoms of different types are illustrated. Different interaction processes of nanostructures and mechanisms of burying the templates and pores were registered for different types of atoms being deposited. Single atoms, which reached the pore bottom, were observed for all types of atoms being deposited. The most complete and dense pore burying was registered during gallium epitaxy.

10.1 SETTING UP AND STAGES OF SOLVING THE PROBLEM OF NANOFILM DEPOSITION

The problem of modeling the formation of nanofilm coatings was solved in several steps. At the first stage, the template from amorphous aluminum oxide was formed. Aluminum and oxygen atoms are put into the computational cell in the required proportion (2:3) with periodic boundary conditions on each side (Fig. 10.1a). The template stabilizes and comes to rest under the action of potential forces under normal thermodynamic conditions (Fig. 10.1b). The template stabilization is conditioned by potential forces, in particular, as it is formed because of the self-organization of aluminum and oxygen atoms. At the same time, heat fluctuations and diffusion are present in the range of the set temperature in the template formed, but there is no essential reconstruction of its structure, the atoms slightly oscillate near the positions they occupy.

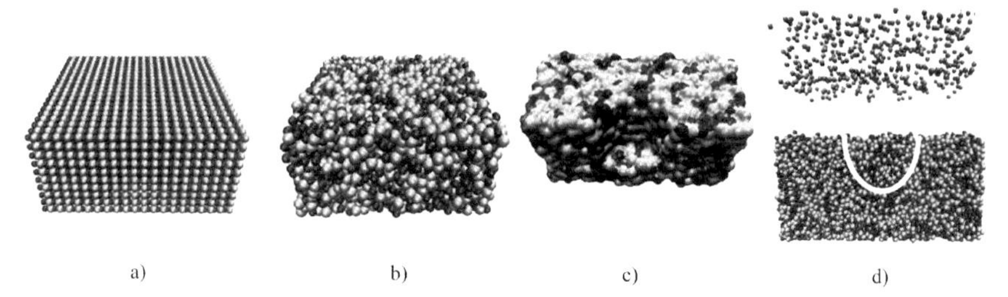

FIGURE 10.1 Steps of solving the problem of forming nanofilm coatings based on porous aluminum oxide: (a) initial state of the system, (b) template relaxation without the pore, (c) template relaxation with the pore, and (d) precipitation.

The hole is cut in the template at the second stage—the pore with the required radius and depth (the cutting of the template with the pore is demonstrated in Fig. 10.1c). Later, this pore will be silted with atoms of different types (Fig. 10.1d). The general pattern of the problem of forming heterogeneous electro-optical coatings is given in Figure 10.1.

The boundary conditions and appearance of the system being modeled are demonstrated in Figure 10.2. Because of the periodic boundary conditions in the directions x and y, only one pore was considered in this work. In horizontal directions, the periodic boundary conditions envisage the parallel transfer of the computational cell. The system being modeled was affected by rigid boundary conditions from the top and bottom. When the atoms were approaching the upper boundary of the system investigated, their bounce from the rigid wall was imitated. The positions of atoms in the thin layer near the boundary of the computational cell were rigidly fixed from the bottom. This type of boundary conditions did not allow the nanosystem atoms to leave the computational area in case of deviation from the main precipitation trajectory.

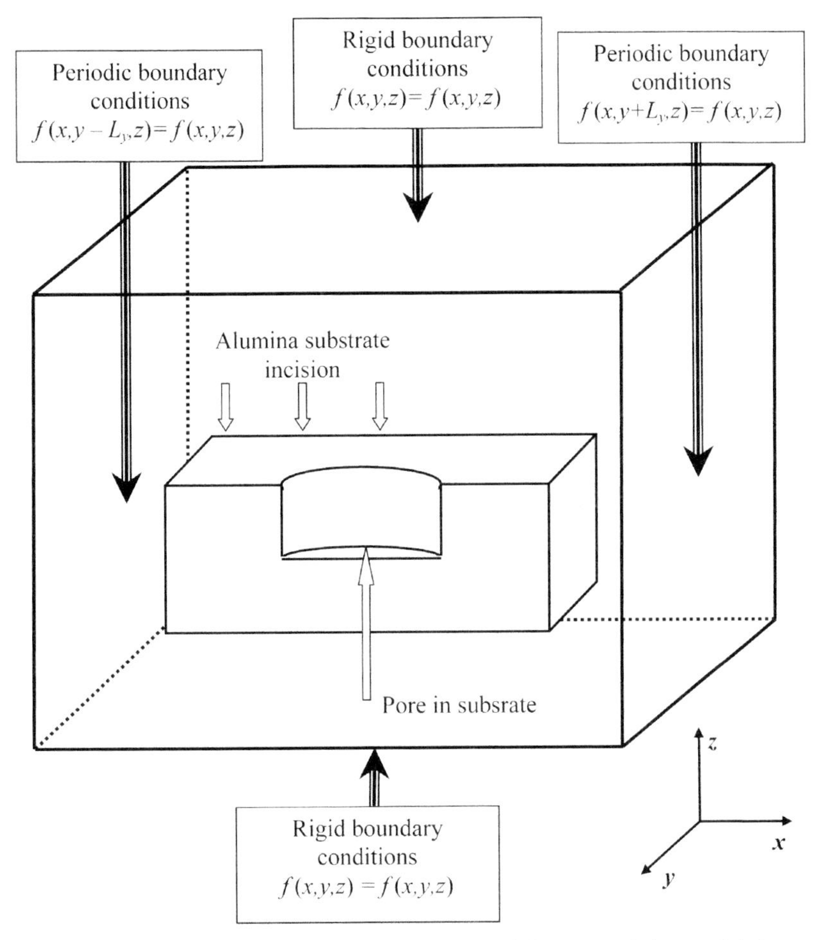

FIGURE 10.2 Boundary conditions and appearance of the system being modeled.

The convergence of numerical solution of the problem set frequently depends on the selection of the corresponding integration step. The step needs to be small enough to correctly reflect the system behavior. When using the methods of molecular dynamics, the value of the mass of substances being modeled influence the integration step value. It is selected in the range from 0.5 up to 2 fs. In this work, the integration step by time is 1 fs. The total time when modeling the system for the stabilization stage (Fig. 10.1a) was about 0.5 ns, for the relaxation stage (Fig. 10.1b,c)—0.2 ns, and precipitation stage (Fig. 10.1d)–0.2 ns.

The porous template was silted by homogenous precipitation of atoms along the normal against the template. The atoms being precipitated were added in the region above the template during the siltation stage. Their position above the template was determined by the uniform random distribution law. The number of atoms added in a time unit and their total number were the process control parameters. The initial velocity of the precipitated atoms was constant. The velocity parameters were changed only under the interaction of the precipitated atoms with the template. To conduct the test calculations, the single nanostructure was considered in the air-free atmosphere and its dynamics during the relaxation self-organization of atoms.

The molecular dynamic modeling was carried out at constant temperature. The constant temperature was maintained in the system with the help of thermostat algorithm. Thermostat is a means of energy extraction and cooling of too fast atoms, as well as a means of energy feeding when the nanosystem is not heated up enough. At present, thermostat algorithms are quite variable: collision thermostat, Berendsen thermostat, friction thermostat, Nose–Hoover thermostat. Nose–Hoover thermostat was used in this work. In the problems of forming nanosized elements, the velocity field at the initial time moment was selected in accordance with Maxwell distribution.

10.2 FORMATION OF MONOFILMS ON POROUS ALUMINA SUBSTRATES

The amorphous aluminum oxide templates with the following dimensions: length—12.4 nm, width—12.4 nm, and height—6.2 nm were used in the modeling process. The total number of atoms in the template after the pore formation was about 60.5 thousand. Before the precipitation process, the template was at rest, at the beginning its temperature was 300 K and it was further maintained at the same level. The graph of the template temperature changes, as well as the kinetic energy for the stabilization and relaxation stages is given in Figure 10.3.

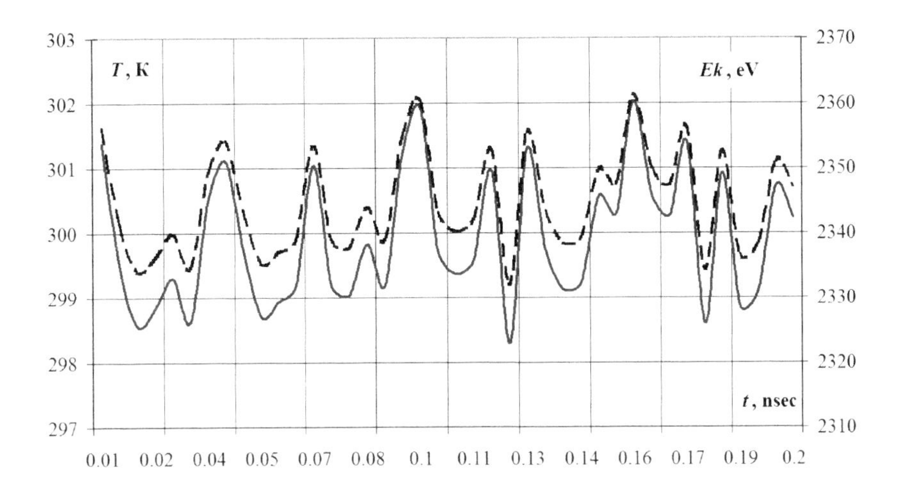

FIGURE 10.3 Dependence of temperature (left ordinate) and kinetic energy (right ordinate) of the porous aluminum template for the relaxation stage.

The pore with the radius of 2 and 4 nm deep was cut in aluminum oxide template. The pore siltation with aluminum oxide was not observed at rest without the atom precipitation. The lower template layer was fixed to avoid its vertical movement at the precipitation stage. The rest of the atoms were not fixed and could freely move in any direction.

Different types of atoms were precipitated onto aluminum oxide template in this work. The number of precipitated atoms was 20,000. The precipitation was uniform along the whole template surface and with the same intensity in time. The atom velocity at epitaxy was 0.05 nm/ps. The physical characteristics of substances used when modeling the process of pore siltation and formation of film coatings are given in Table 10.1.

TABLE 10.1 Physical Properties of the Precipitated Elements.

Symbol	Name	Standard atomic weight, amu	Crystal lattice structure	The lattice parameter, nm	Melting temperature, K
Au	Gold	196.967	face-centered cubic 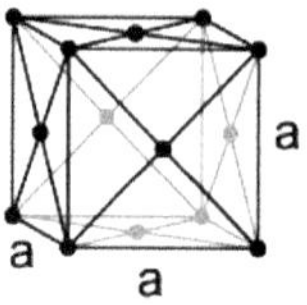	a = 0.4078	1337.33
Ag	Silver	107.868	face-centered cubic 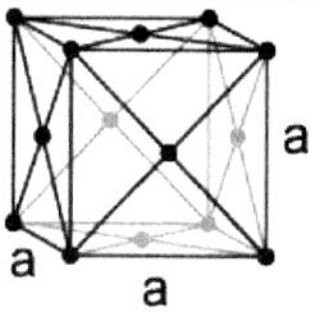	a = 0.4086	1235
Cr	Chromium	51.996	body-centered cubic 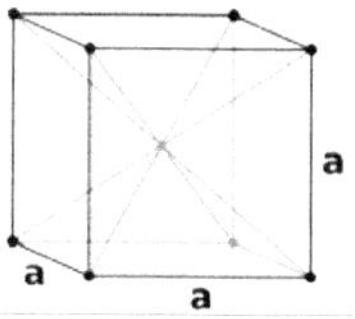	a = 0.2885	2130
Cu	Copper	63.546	face-centered cubic 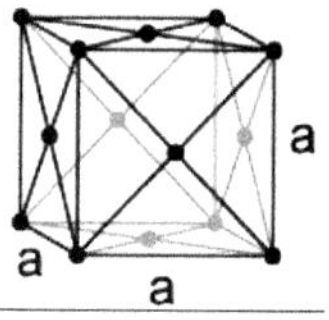	a = 0.3615	1356
Fe	Iron	55.847	body-centered cubic 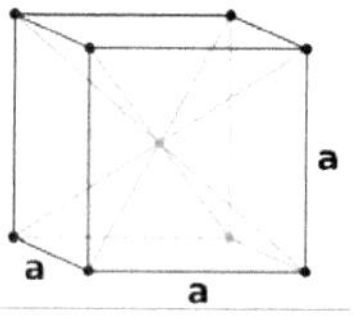	a = 0.2866	1812

TABLE 10.1 *(Continued)*

Symbol	Name	Standard atomic weight, amu	Crystal lattice structure	The lattice parameter, nm	Melting temperature, К
Ga	Gallium	69.723	Orthorhombic 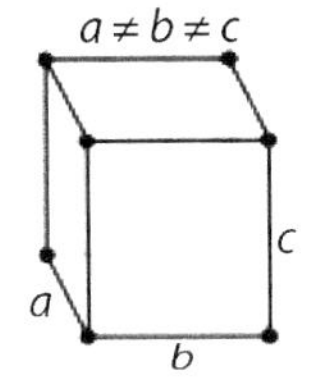	a = 0.4519 b = 0.7658 c = 0.4526	302.93
Ge	Germanium	72.630	face-centered diamond-cubic 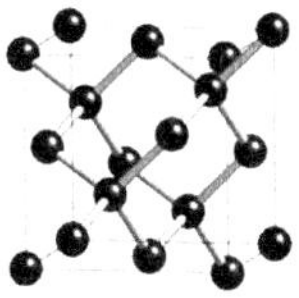	a = 0.566	1210.6
Ti	Titanium	47.867	hexagonal close-packed 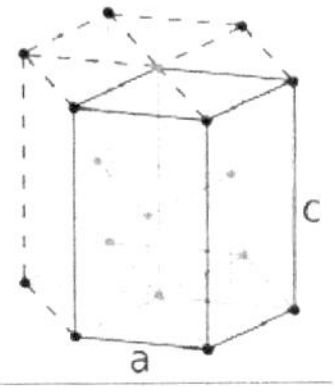	a = 0.2951 c = 0.4697	1933 ± 20
Pd	Palladium	106.42	face-centered cubic 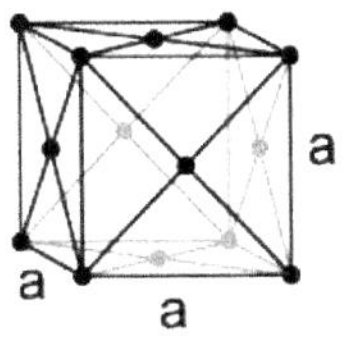	a = 0.3890	1827
Pt	Platinum	195.084	face-centered cubic 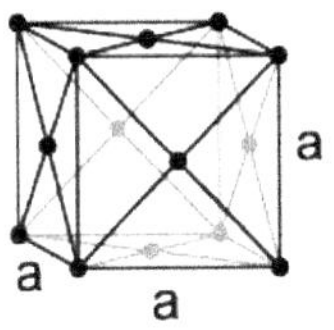	a = 0.3920	2041.4

TABLE 10.1 *(Continued)*

Symbol	Name	Standard atomic weight, amu	Crystal lattice structure	The lattice parameter, nm	Melting temperature, K
V	Vanadium	50.9415	body-centered cubic	a = 0.3024	2160

The results of epitaxial nanofilm formation from silver atoms on the template of porous aluminum oxide are given in Figure 10.4. The atom precipitation was uniform, the formation of large agglomerates in air environment was not observed. The film formed on the template was uniform with little sinking in the pore region.

The pore was not completely silted with silver atoms; it was observed that, part of silver atoms got inside the pore near its upper part. The rest of the pore was hollow during the entire epitaxy stage. The template central layer 0.2 nm thick with silver nanofilm formed on it is shown in Figure 10.4 at the right. The figure analysis confirms the incomplete pore siltation.

There are practically no silver atoms left in the air environment above the template by the time moment of 0.2 ns of precipitation stage. The partial pulling of oxygen atoms out of the template and their getting into the nanofilm lower layers on the template surface are observed. The slight migration of oxygen and aluminum atoms during the whole modeling stage takes place, which can be explained by the temperature movements of the system atoms.

The horizontal section of nanofilm coating along the template surface after the precipitation of silver atoms for the precipitation time of 0.2 ns is demonstrated in Figure 10.5. The partial formation of crystalline structure formed by aluminum atoms is observed in some parts.

Silver atoms in the template center in Figure 10.5 are only on the surface, the pore is hollow inside. The picture asymmetry is explained by pseudostochastic behavior of the nanosystem produced by the temperature laws of initial distribution of atom velocities.

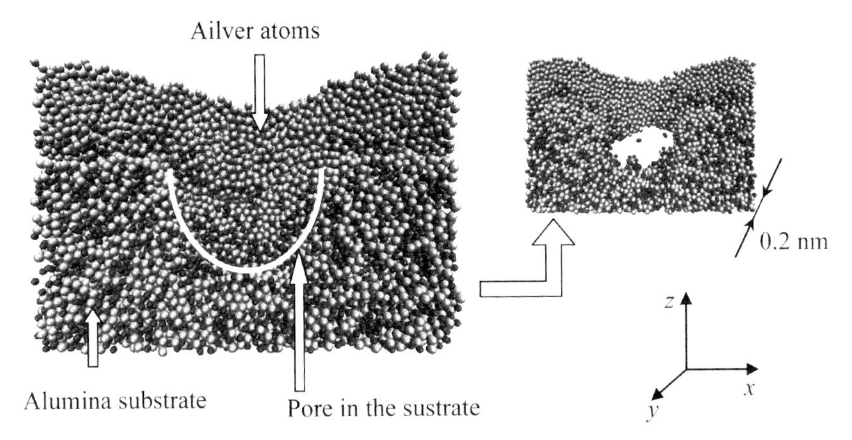

FIGURE 10.4 Results of silver atom precipitation onto the template of porous aluminum oxide, precipitation time—0.2 ns.

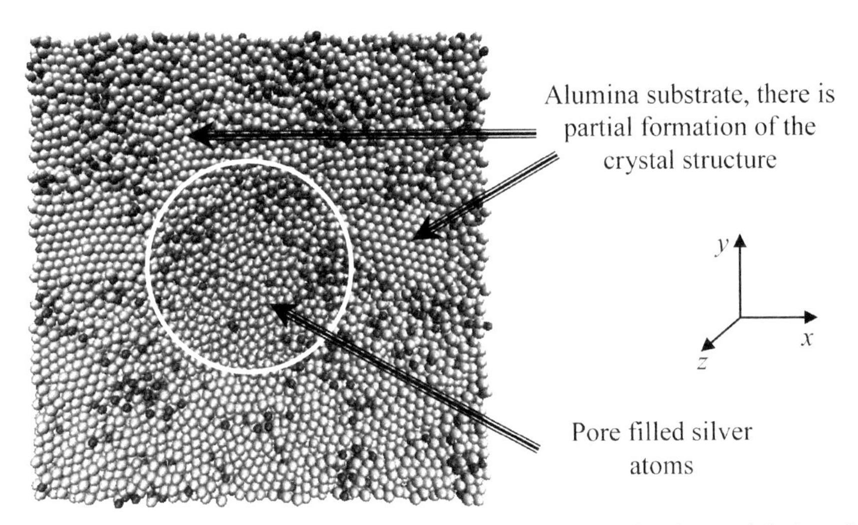

FIGURE 10.5 Horizontal section of nanofilm coating surface after the precipitation of silver atoms along the template surface, precipitation time—0.2 ns.

The results of gold and magnesium atom precipitation onto the template of porous aluminum oxide are similar to the previously described process of sliver atom epitaxy. In such cases, the template is coated quite uniformly with bending in the pore position. At the same time, the pore is not silted completely; the precipitated atoms produce something like a cork or cover near its surface.

In the process of template siltation, chromium, copper, and iron atoms start grouping into nanostructures in the air, before reaching the template surface. The significant coarsening of agglomerates is not observed, they continue moving to the template. The example of iron atom precipitation is shown in Figures 10.6 and 10.7.

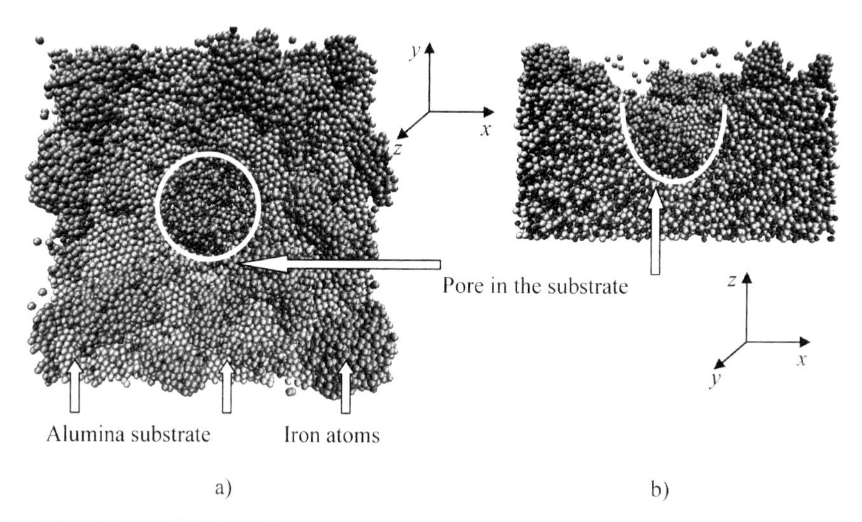

FIGURE 10.6 Siltation of the template of porous aluminum oxide with iron atoms: (a) top view and (b) vertical section along the pore center, precipitation time—0.05 ns.

The siltation of the template of porous aluminum oxide with iron atoms for the precipitation time of 0.05 ns is demonstrated in Figure 10.6. As iron atoms started grouping in the air environment above the template already, the template was silted following the island principle. Small iron nanostructures were gradually growing on the template and grouping into the bigger ones. The formation of one of iron nanostructures was observed inside the pore, it is especially vividly seen on the vertical section along the pore center (Figure 10.6b). Oxygen atoms from the template upper layers are actively interacting with iron atoms forming amorphous oxide structures.

Figure 10.7 (axis z is inclined by 45° to the observer) illustrates that the surface of iron nanofilm formed on the template is uneven.

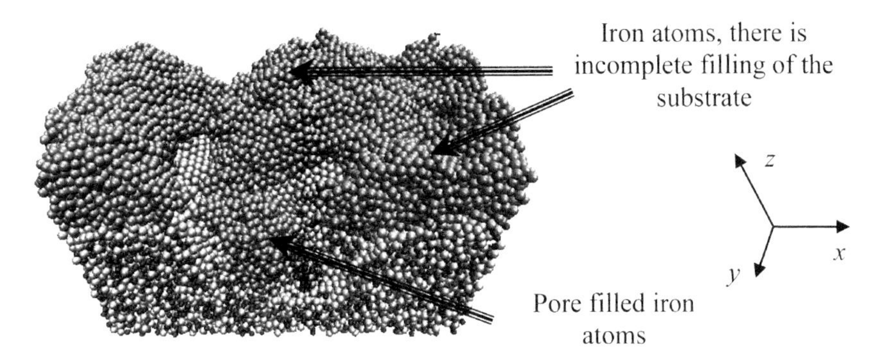

FIGURE 10.7 Siltation of the template of porous aluminum oxide with iron atoms: vertical section along the pore center, precipitation time—0.2 ns.

In contrast to the epitaxy of silver and gold atoms, iron does not completely cover the template. Besides, rather significant height differences, of several nanometers sometimes, are observed in the nanofilm. The same results as for iron atoms are characteristic for the coatings of porous templates with chromium atoms, except that the nanofilm is even and chromium atom agglomeration in the air environment is less intensive.

The precipitation of gallium and germanium atoms is similar to each other in the physical process. The result for the time of 0.2 ns is given in Figure 10.8.

The pore in these cases is also not silted completely. The nanofilm is formed by regions on the template surface; the nonsilted template is also seen as large regions in Figure 10.8. Small gallium nanoparticles are seen on the template surface. Oxygen atoms are more intensely pulled out of the upper layers of the template than in the previously considered processes.

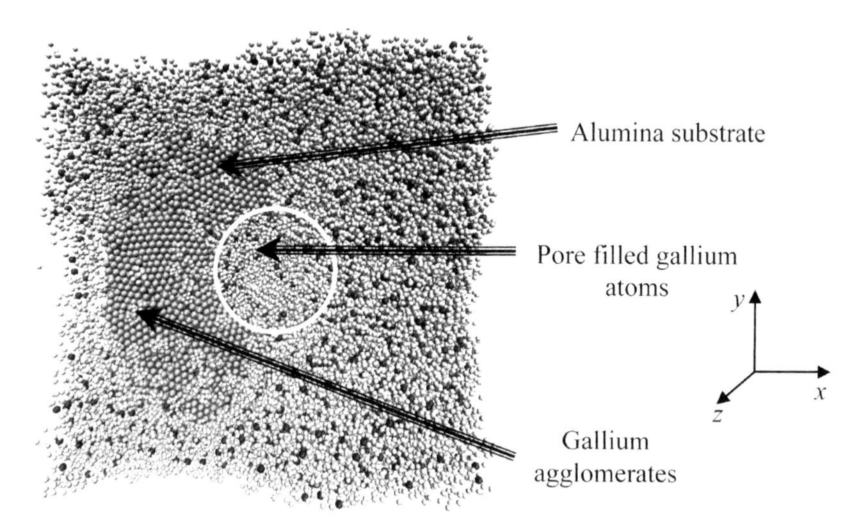

FIGURE 10.8 Siltation of the template of porous aluminum oxide with gallium atoms, precipitation time—0.2 ns.

When precipitated, germanium atoms more intensely penetrated the pore region, but further in the process of nanofilm formation they were pulled out onto the surface. The surface profile of the obtained nanofilm from germanium atoms was not uniform with height differences and large nonsilted regions.

The interesting effect was observed during the epitaxy of palladium and platinum atoms onto the template of porous aluminum oxide. The precipitation result for palladium atoms in two projections is given in Figure 10.9. In this case, the uniform nanofilm was formed with slight sinking in the pore region. However, the nonsilted opening was observed immediately above the pore during the whole condensation stage. Palladium atoms got inside the pore only insignificantly, as it is seen from the vertical section along the pore center demonstrated in Figure 10.9b.

The siltation of the template of porous aluminum oxide with titanium atoms for the precipitation time of 0.2 ns is demonstrated in Figure 10.10. For this type of atoms, the nanofilm formed looked incoherent and coarse. The atoms got inside the pore only insignificantly and stayed on the template surface during the whole condensation stage.

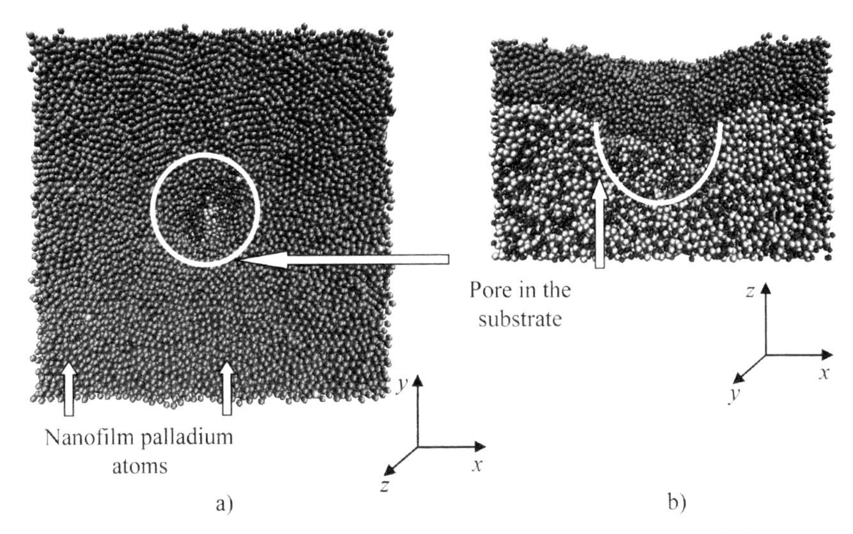

FIGURE 10.9 Siltation of the template of porous aluminum oxide with palladium atoms: (a) top view and (b) vertical section along the pore center, precipitation time—0.2 ns.

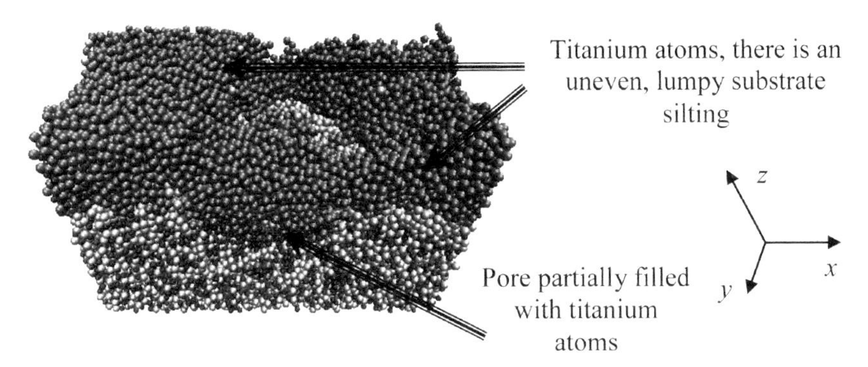

FIGURE 10.10 Siltation of the template of porous aluminum oxide with titanium atoms: vertical section along the pore center, precipitation time—0.2 ns.

The siltation of the template was observed for its epitaxy with vanadium atoms. A large number of vanadium atoms got inside the pore, but nonsilted regions remained on the template surface itself. The translation results of the periodic computational cell relatively to the perpendicular of the precipitation plane during the epitaxy of iron atoms are shown in Figure 10.11.

The graphs of penetration of considered substance atoms into the pore on aluminum oxide template are shown in Figure 10.12. The most penetration was demonstrated by gallium atoms, the least by palladium ones. Gold demonstrated the best penetration among noble metals. After about 120 ps, the intensity of all atom penetration goes down and further it changes insignificantly that is seen in Figure 10.12. Iron and gallium atoms demonstrated the highest intensity at the initial modeling stage (approximately up to 90 ps).

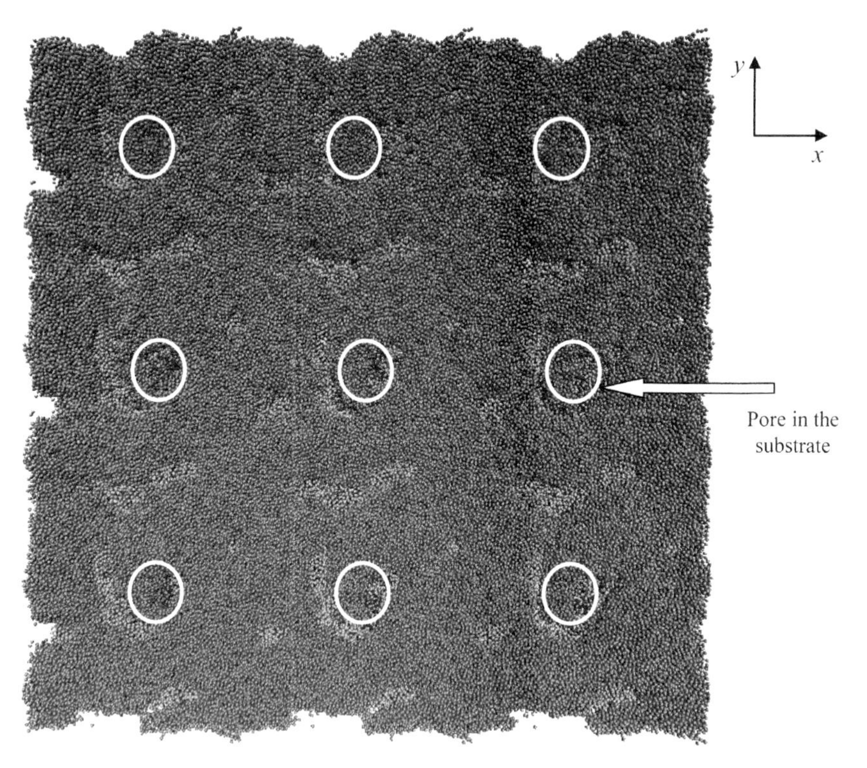

FIGURE 10.11 Siltation of the template of porous aluminum oxide with iron atoms: top view, precipitation time—0.05 ns.

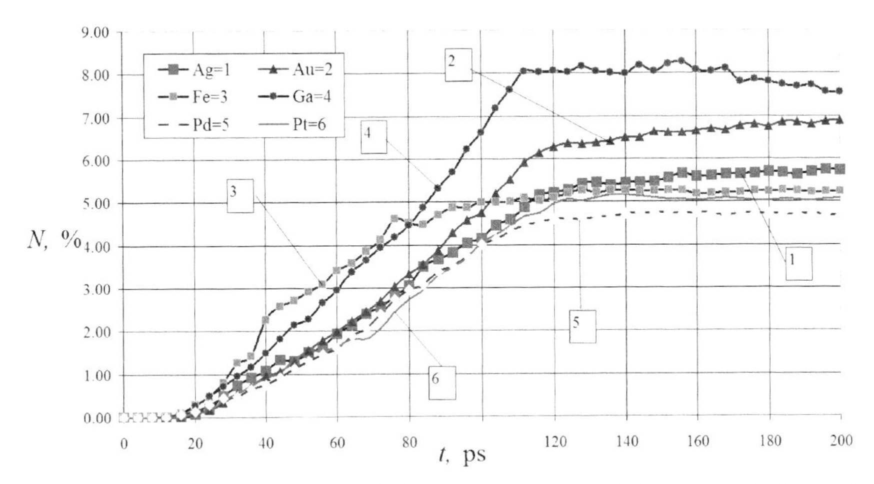

FIGURE 10.12 Percentage of precipitated atoms that got into the pore to the total number of precipitated atoms.

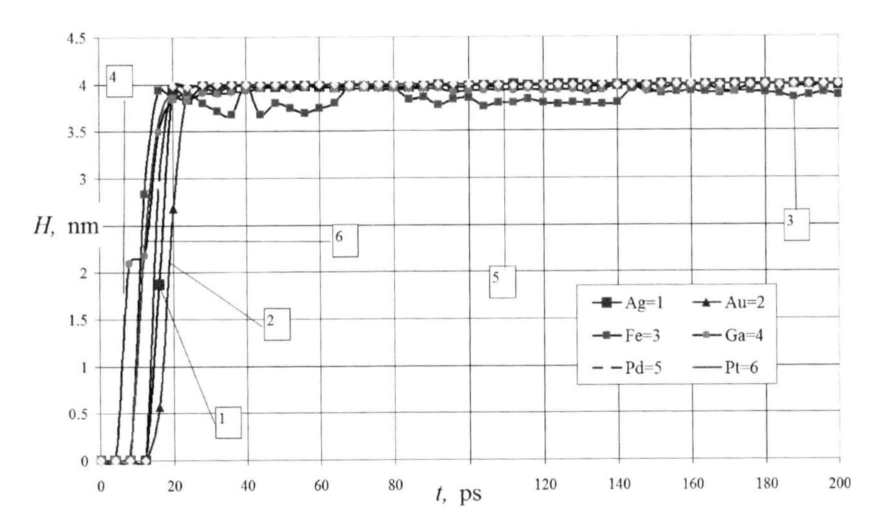

FIGURE 10.13 Penetration depth into the pore of precipitated atoms.

Since the initial distribution of atoms has a stochastic character, it would be interesting to see what considered elements will penetrate the pore to the maximum depth. From the graph in Figure 10.13, it is seen that the considered atom types completely penetrate the pore at the initial stage (to the pore height). And the penetration intensity at the initial stage (up to 20 ps) is similar.

The dynamics of the pore filling with precipitated atoms is given in Figure 10.14, where you can see the graphs of the depth of mass center of atoms, which penetrated the pore.

The depth of mass center is calculated only relatively to the atoms, which filled the pore. Therefore, at the initial time moments (up to 20 ps), the greater shifting of the mass center is observed. Furthermore, the number of atoms penetrating the pore grows, the dependencies in Figure 10.14 shift in the direction of the middle of the pore depth. The shift of the mass center reaches the most significant depth when iron and gallium atoms are precipitated.

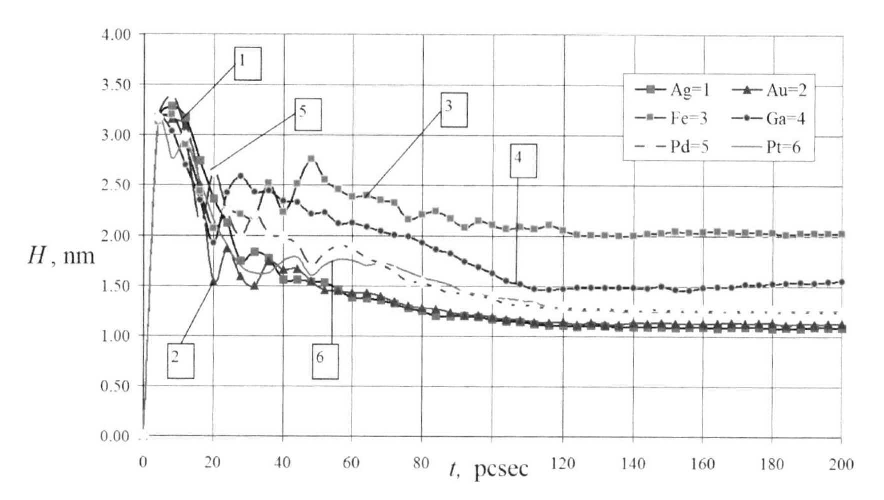

FIGURE 10.14 Depth of mass center of atoms, which penetrated the pore.

10.3 EFFECT OF PORE SIZE INTO SUBSTRATES ON THE NATURE OF THE DEPOSITED NANOFILM

For the following series of computational experiments, the pore radius in the template varied, the depth being the same (4 nm). Gallium atoms were used for precipitation as one of the most suitable to form nanostructured objects on the template. The graphs of Ga atoms, which got inside the pore in percent relatively to the total number of precipitated atoms, are demonstrated in Figure 10.15. Analyzing the graphs in Figure 10.16, you can see that the number of atoms actively grows in the time interval of 20–120 ps. The pore siltation after 120 ps of condensation is followed by the reconstruction of atomic structure that corresponds to the stabilization of dependencies and slight decrease in the percentage of atoms, which penetrated the pore.

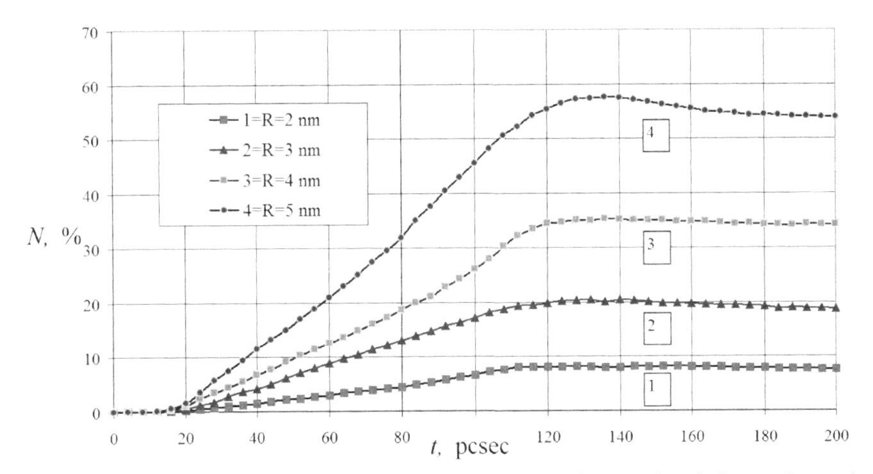

FIGURE 10.15 Percentage of Ga atoms, which got into the pore in relation to the total number of precipitated atoms.

Besides, during the siltation of pores of different radii with gallium atoms, the mass center of precipitated atoms stabilizes at different depths of the pore. For the pores with the radii of 2 and 3 nm, the mass center is formed above the middle of the pore depth. With the size growth, the mass center starts to be formed in one place—near the middle of the pore depth. This fact allows saying that the further radius growth (over 5 nm) will not significantly influence the mass center, the pore is already rather compactly filled with the precipitated atoms.

The processes of formation of nanofilms and self-organization of the deposited atoms can be detailed through a change in the energy of these atoms. The dependence of the potential energy E and the total energy E_{tot} of epitaxial silver atoms depending on the temperature are shown in Figure 10.17. Measurements of energies and temperatures were carried out through equal time steps of 4 ps. The temperature of the deposited atoms gradually decreased. The process of condensation took place from the heated state. The direction of the condensation process is shown shaped arrow in Figure 10.17.

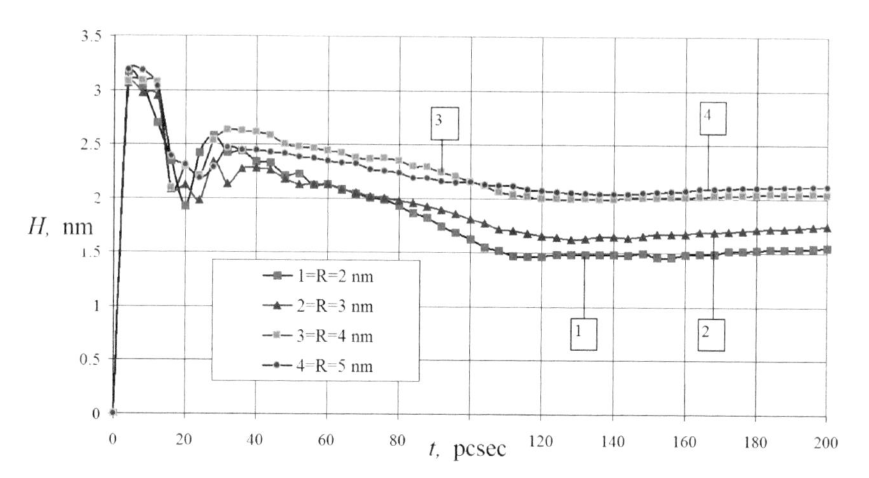

FIGURE 10.16 Depth of mass center of Ga atoms, which penetrated the pore.

The deposited silver atoms were in a free state at the initial instants of time. Their speed was directed to the substrate, so they had increased kinetic energy. The thermostat was not applied to condensable atoms. Their energy was reduced by interacting with a substrate that had a lower temperature. Excess energy was removed from the substrate to match the temperature $T = 300K$.

The cooling of the deposited silver atoms did not occur uniformly, as can be seen from the graphs in Figure 10.17. The region of the phase transition is highlighted in a rectangular frame. The region corresponds to a time moment of 92–100 ps from the beginning of the condensation stage. The rebuilding of the structure into an energetically more favorable state took place approximately 8 ps. At the same time, the energy changed and the temperature remained at the same level. Potential and total energies behave on the charts in a similar way.

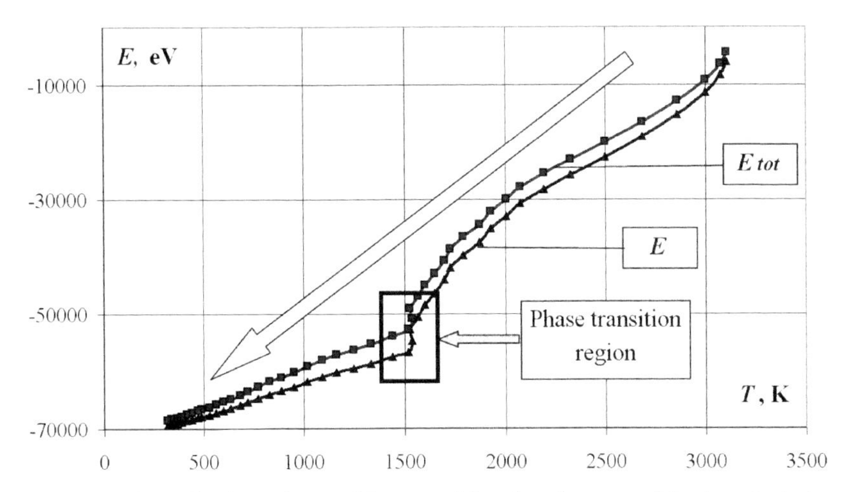

FIGURE 10.17 The dependence of the potential and total energy of the deposited atoms on the condensation stage.

The melting point of silver is 1235 K, the boiling point is 2485 K. The dynamics of the formation of the nanofilm shows that the silver atoms are not separately collected into droplets and nanostructures. Consequently, the phase transition occurs mixed at a temperature between the melting and boiling of silver $T = 1520$K.

10.4 INVESTIGATION OF THE STRUCTURE OF THE TEMPLATES AND NANOFILMS FORMED

The material structure both for porous and solid templates was investigated in this work (Fig. 10.18). Later, the template is buried with the deposited atoms.

Parameter C found by the following formula is applied to determine the material structure[1,2]:

$$C = \sum_{i=1}^{Z/2} \left| r_i + r_{i+Z/2} \right|^2, \tag{10.1}$$

where Z = number of the current neighbors for the current atom;

r_i and $r_{i+Z/2}$ = radius vectors from the central to the pair of closely located atoms.

Z(Z–1)/2 possible pairs of the atom nearest neighbors are considered in eq 10.1. Because of the symmetry of crystalline lattice, this parameter tends to zero for ideal crystals, and it will have a positive value for amorphous ones.

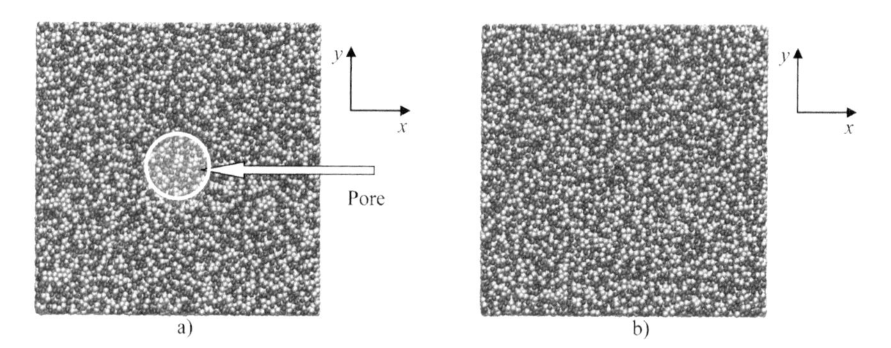

FIGURE 10.18 Exterior view of porous (a) and solid (b) template of aluminum oxide.

The templates of aluminum oxide with the following dimensions: length—15.3 nm, width—15.3 nm, height—7.6 nm were used during modeling. The total number of atoms in the template after the pore formation was approximately 62,000. Before the deposition process, the template was at rest, its temperature was 293 K and it was further maintained at the same level. In this work, the atoms were deposited onto solid and porous templates of aluminum oxide. In case of porous templates, the dimensions of a blind-ended pore were as follows: radius—2 nm, depth—4 nm. The template lower layer was fixed to prevent vertical movement during the deposition stage. The rest of the template atoms were not fixed and could freely move in any direction.

The templates were buried with zinc and sulfur atoms in equal proportions—40,000 atoms of each type. The deposition was uniform along the whole template surface and the intensity by time was equal. The atom velocity during epitaxy was 0.05 nm/ps. In some computational experiments, 5% of copper was added to the atoms being deposited. Similar epitaxial compositions are conditioned by the fact that the investigation is related to specific technological processes used in practice when producing samples with unique optical properties.

The results of epitaxial formation of the nanofilm from zinc and sulfur atoms on the template of porous aluminum oxide are given in Figure 10.19. The height of the nanofilm obtained was 7.2 nm. The deposited atoms formed the non-uniform relief with irregularities observed on the surface. The pore in the template was not buried, but zinc and sulfur atoms partly got into it (Figure 10.19b). In Figure 10.19a, the hole with partial plugging is seen immediately above the pore. The deposition picture was similar for solid templates of aluminum oxide.

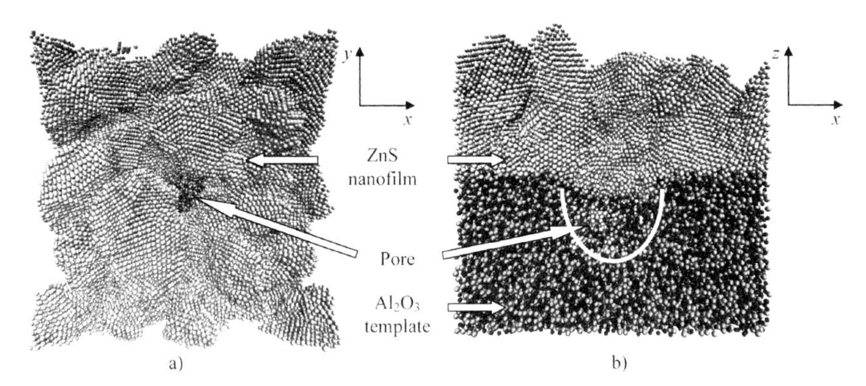

FIGURE 10.19 Results of deposition of Zn and S atoms onto the template of porous aluminum, deposition time—0.5 ns.

We can judge about the structure (crystalline or amorphous) of the materials by the parameter of crystalline lattice ideality calculated in eq 10.1.

This parameter is calculated for the whole group of atoms, then its average value if found. The values of the parameter of crystalline lattice ideality for the template of aluminum oxide are given in Figure 10.20.

Four cases of computational experiment were considered in this work: solid template with the deposition of Zn and S atoms; solid template with the deposition of Zn, S atoms, and 5 % of Cu atoms; porous template with the deposition of Zn and S atoms; porous template with the deposition of Zn, S atoms, and 5 % of Cu atoms. Besides, the structure of both the template and nanofilm formed was evaluated.

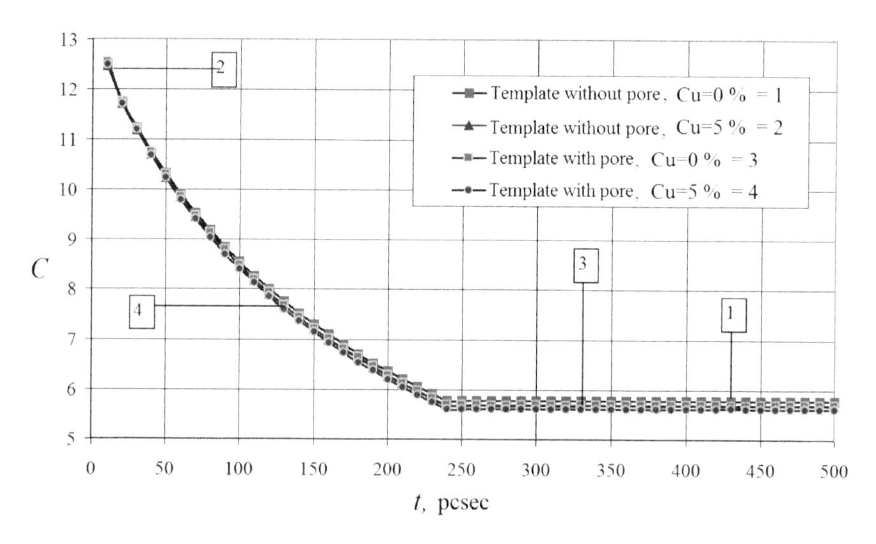

FIGURE 10.20 Average parameter of the crystalline lattice structure for the template.

Figure 10.21 demonstrates the change in the structure of the template crystalline lattice at the stage of the structure rearrangement (magnified region of Fig. 10.20). The analysis of dependencies of Figures 10.20 and 10.21 shows that the parameter of crystalline lattice ideality for all four cases is similar. The time of 250 ps corresponds to the moment when the deposition of atoms stops, the rearrangement of their coordinates into more energy beneficial position starts.

The template practically does not change its structure at the time moments of 250–500 ps, only the insignificant temperature fluctuations of atoms near the crystalline lattice nodes are observed. The value of C parameter for the template is rather large that indicates the material amorphous structure. From Figure 10.21, it is seen that the structure parameter value for the investigations carried out is slightly different.

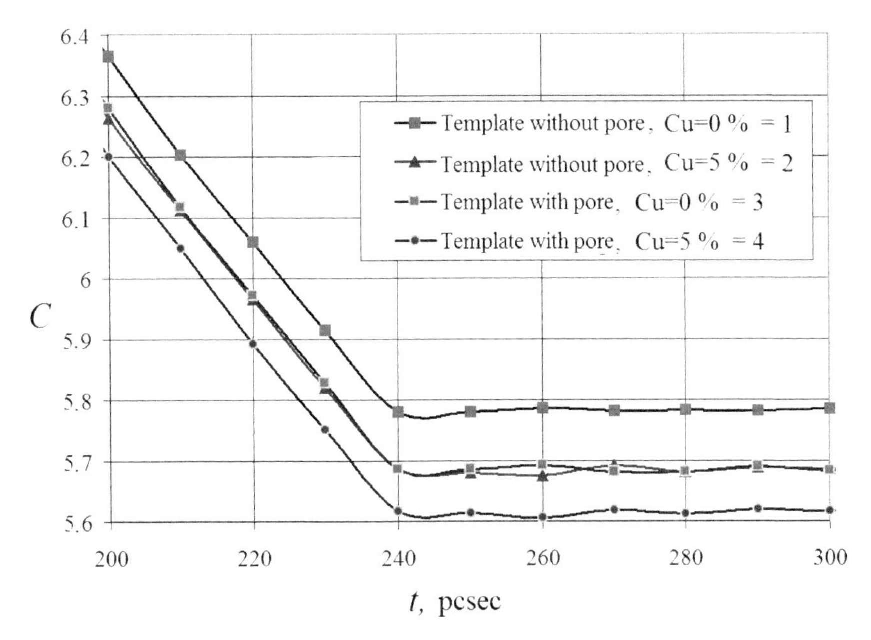

FIGURE 10.21 Average parameter of the crystalline lattice structure for the template at the stage of structure rearrangement.

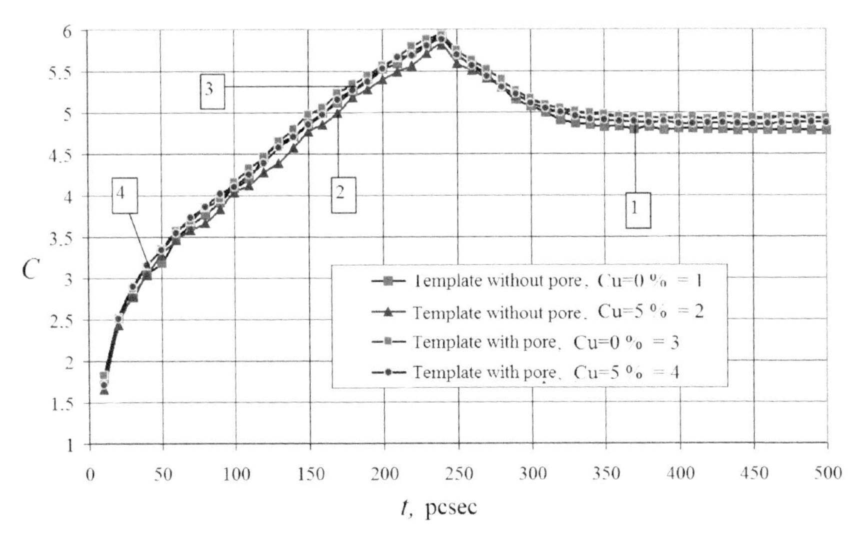

FIGURE 10.22 Average parameter of the crystalline lattice structure for the atoms being deposited.

For all four variants of computational experiments, the structure parameter change was also calculated for the atoms being deposited (Figs. 10.22 and 10.23). The distributions of these parameters are characterized by active rearrangement of atom positions during evaporation and following only slightly varying structure. The magnified region of the deposition completion (Fig. 10.23) allows seeing the differences in the properties of the materials being formed. During the condensation, the least value of the crystalline structure parameter is observed in the computational experiment without a pore with the addition of 5% of copper atoms. The stabilized evaporated material has the least parameter value for the template without a pore and without the addition of copper atoms. The amorphous behavior of the template and deposited nanofilms is characteristic for all the above cases.

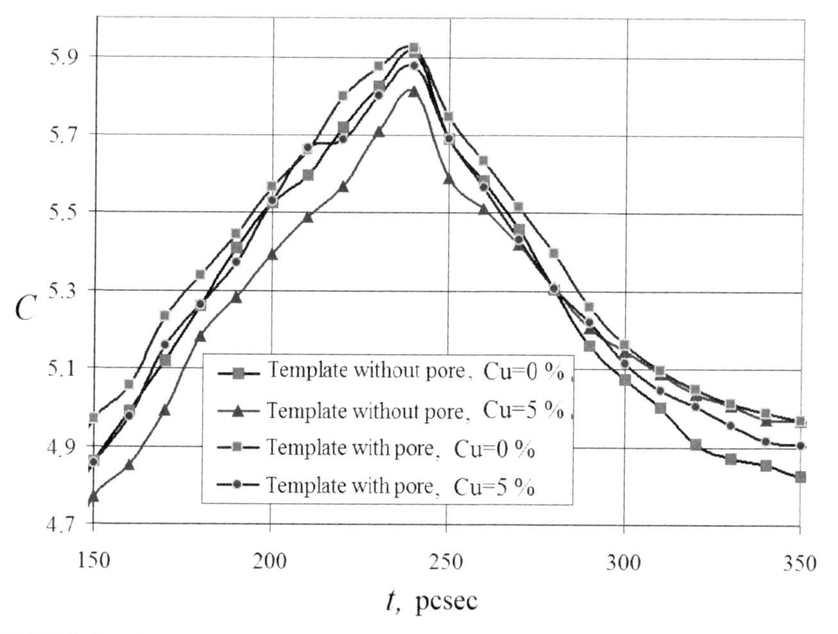

FIGURE 10.23 Average parameter of the crystalline lattice structure for the atoms being deposited at the structure rearrangement stage.

10.5 FORMATION OF MULTICOMPONENT NANOFILMS AND FILMS WITH ADMIXTURES ON THE TEMPLATES OF POROUS ALUMINUM OXIDE

In continuation of the investigations of nanofilm formation on porous templates, the deposition of molecular ZnS was considered in this work. In multiparticle potentials, such as Abel–Tersoff, Stillinger–Weber, embedded atom method, the bond formation and breakage occur because of the principles of force field performance. Nevertheless, for multiparticle potentials, rather large amount of empirical parameters is required. Especially for alloys and complex compounds, it is necessary to obtain additional characteristics of force fields, adapting the action of energy fields for multimolecular nanostructures. Empirical constants are not always known for complex compounds; therefore, in some approximation, we have to content ourselves with simpler types of potentials of molecular dynamics. In consequence of the aforementioned, Lennard–Jones potential with additionally introduced account of the nanosystem bound atoms was used for modeling ZnS deposition.

Nanofilms based on ZnS are actively used in optical systems of IR band. For industrial purposes, ZnS is produced by chemical deposition from zinc gases and vapors onto the template. The material is characterized by high ultimate strength, absolute mechanical hardness, and chemical inertness that make it irreplaceable for military applications and other severe operating conditions. ZnS is a semiconducting material with energy bandgap of 3.54–3.91 eV and it is also used in semiconductor lasers, medicine, optical electronics, and display production.

ZnS is not always applied in the pure state, the material is frequently added to admixtures and additives. The introduction of silver admixture into the composition results in the luminescence in the blue light region. The addition of copper as an alloying metal allows using the luminescence of green color, which is applied in display boards, panels, luminophors, and oscillograph tubes.

The manufacturing technology of electroluminophors with yellow luminescence is mainly based on the synthesis of solid solutions of zinc and cadmium chalcogenides activated by copper. This class of luminophors comprises zinc sulfoselenide Zn(S,Se):Cu and zinc cadmium sulfoselenide (Zn,Cd)(S,Se):Cu. However, the application of these grades of luminophors is limited by the application of cadmium and selenium compounds in the process of their manufacturing, which are harmful for the ecology when getting into the environment.

Apart from sulfoselenide electroluminophors, radiating in the yellow spectrum band, there is also electroluminophor based on ZnS activated by copper and manganese (ZnS:Cu,Mn). The advantage of this composition against sulfoselenide one is obvious since toxic compounds of cadmium and selenium are not used during its manufacturing. However, low luminescence brightness and high brightness degradation rate require the technology improvement. Identification of the reasons determining the brightness increase of ZnS:Cu,Mn at the changed synthesis conditions is of high priority.

When investigating luminophor properties, the luminescence brightness is usually the determining parameter. However, to find out main physical mechanisms underlying luminescence, it is necessary to investigate both spectral and kinetic characteristics of electroluminophor that confirms again the importance of theoretical studies of processes of nanofilm deposition onto porous templates.

In order to obtain realistic results maximally fitting the experimental data, the sizes of the modeling region were significantly increased. The amorphous aluminum oxide templates with the following dimensions: length 19.1 nm, width 19.1 nm, and height 11.6 nm were used in the modeling process. The total number of atoms in the template after the pore formation was about 122,000. Before the precipitation process, the template was at rest, at the beginning its temperature was 293 K and it was further maintained at the same level.

The pore with the radius of 5 and 10 nm deep was cut in aluminum oxide template. The pore siltation with aluminum oxide was not observed at rest without the atom precipitation. The lower template layer was fixed to avoid its vertical movement at the precipitation stage. The rest of the atoms were not fixed and could freely move in any direction.

The number of precipitated ZnS molecules was 200,000. The proportion of alloying elements increased in proportion to its composition increased the percentage of epitaxial atoms. The precipitation was uniform along the whole template surface and with the same intensity in time. The atom velocity at epitaxy was 0.05 nm/ps.

The evaporation process of nanofilms from pure ZnS is demonstrated in Figure 10.24. The duration of the complete deposition stage was 0.6 ns. Analysis of the graphical results indicates that a pore is gradually buried with nanofilm. Initially, the neck starts forming on the sides above the hole (Figure 10.24a), which is gradually covered later on. ZnS molecules partially get into the pore, but its complete dense filling does not occur (Figs. 10.24b,c). Nevertheless, practically all internal surface of the porous hole appears to be covered with ZnS molecules by the deposition stage completion. The gradual pore filling results in the emergence of rounded overgrowths above the pore region.

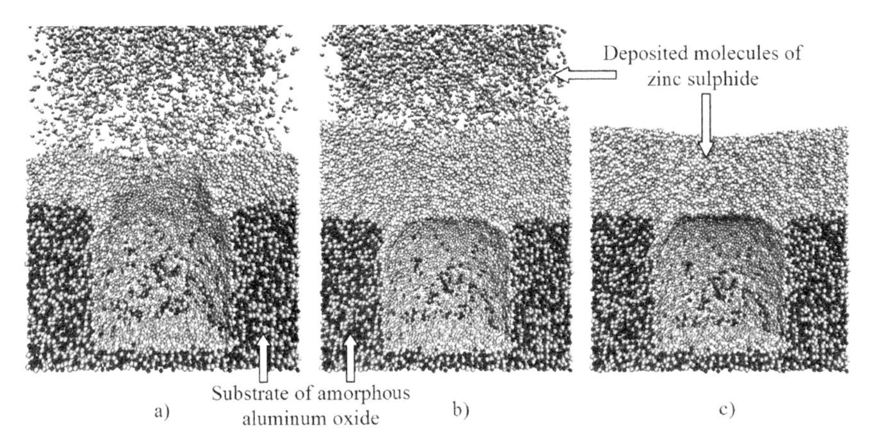

FIGURE 10.24 Result of burying the porous template of aluminum oxide with zinc sulfide for the deposition time: (a) 0.2 ns, (b) 0.4 ns, and (c) 0.6 ns.

In general, the surface of ZnS nanofilm formed is rather even with slight flash above the pore region. The formation of molecular agglomerates above the template during epitaxy is not registered; therefore, the resultant film does not have considerable relief changes in the surface. The nanofilm intensity growth was even. The resultant thickness of the nanofilm formed for pure ZnS was 6.6–6.8 nm.

The process on nanoformation growth in the pore of aluminum oxide template is shown in Figure 10.25. The percentage ratio of nanostructures formed by the deposited atoms was found as applied to the total amount of epitaxially dusted particles. As seen from Figure 10.25, the introduction of alloying elements into ZnS nanofilm, which were uniformly placed in the deposition region, does not significantly influence the pore burying process. Here and further, the percentage of additives is given in the graphs in brackets after the indication of their chemical formulas.

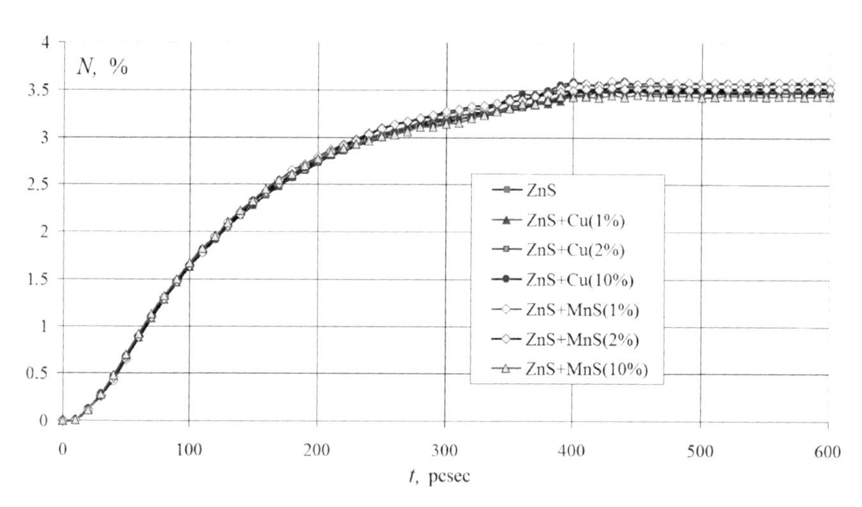

FIGURE 10.25 Percentage of deposited atoms, which got into the pore to the total number of the deposited atoms.

In general, the burying process can be formally divided into two stages. At the first stage (0–400 ps), the cavity in the template is partially filled. The nanoformations grow evenly without sudden jumps and shifts that is confirmed by the dependencies in Figure 10.25. The second stage is characterized by the stationary behavior of lines and time moments of 400–600 ps. At this stage, the specific cover is formed above the pore and newly deposited atoms do not already get into the cavity. Both the visual observation and dependencies in Figure 10.25 indicate that when the alloying elements are added to the nanofilm being formed, there are no fundamental changes in the growth processes, only a slight deviation of functional curves in Figure 10.25 within not more than 0.3% is observed.

Analysis of the graphs of penetration depth of deposited atoms into the pore given in Figure 10.26 demonstrates that for all cases of computational experiments, some atoms practically immediately reach the pore maximum depth of 10 nm. As seen in Figure 10.24, the layer of epitaxially deposited atoms, which contribute accordingly to the depth graphs given in Figure 10.26, is formed on the bottom of the porous hole.

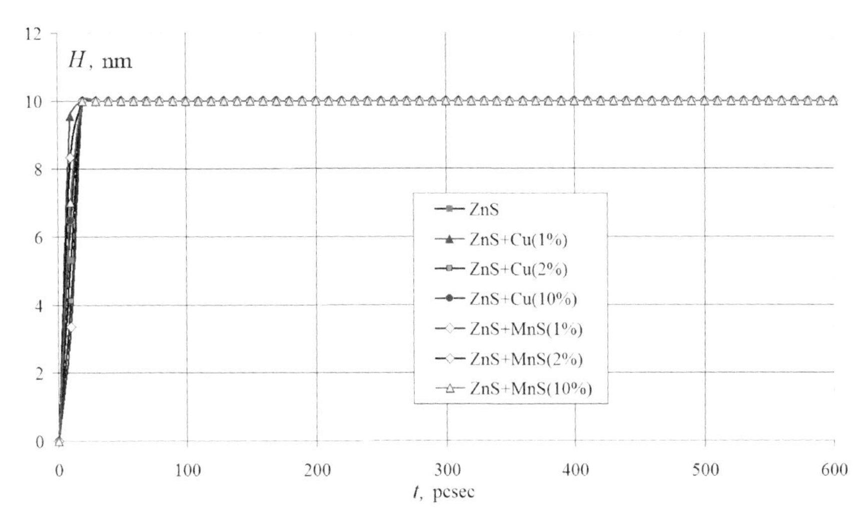

FIGURE 10.26 Penetration depth of the deposited atoms of zinc sulfide with alloying additives into the pore.

The level of filling the porous formation in the template and processes of nanostructure growth can be characterized with the help of investigation of the mass center position of nano-objects formed inside the pore. Location of the mass center of nanoformations inside the pore was found against its bottom and not the template surface (see the right side of Fig. 10.27).

For more convenience, the value of the mass center height was made dimensionless, dividing by the pore depth, and the dependencies obtained are given in percentage ratio.

For the time moments of 0–50 ps, when nanostructures inside the pore only start being formed, the significant increase in the mass center height is observed.

The deposited atoms, which initially got onto the pore bottom, were supplemented by particles on the template cavity walls that resulted in the mass center shift. Later (time moment of 50–600 ps), the mass center position changed insignificantly, the determination of the mean coordinate slightly under the pore depth (level of 45%) was registered.

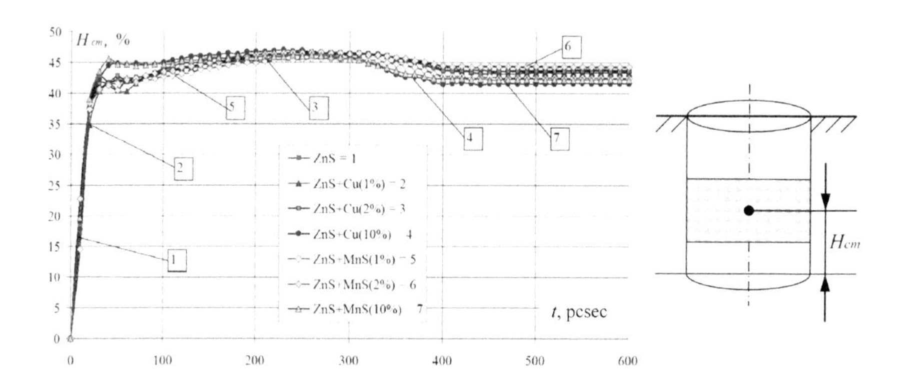

FIGURE 10.27 Relative depth of the mass center of nanoformations in the pore for different alloying additives.

The value, relief, and structure of the nanofilm being formed on the surface play an important role. Growth of the nanofilm thickness on the template surface, taking into account different alloying additives, is demonstrated in Figure 10.28. The averaged value of the nanofilm height above the whole template surface was taken as its average thickness (see the right side of Figure 10.28). The initial time moments of the condensation stage (0–400 ps) are characterized by practically linear increase in the nanofilm thickness. At the time moments of 400–600 ps, the deposited atoms and molecules are compacted and the internal structure is rearranged, so, the film average thickness slightly decreases. The difference in nanofilm values for different types and amounts of alloying additives is observed in Figure 10.28. The introduction of additives results in the increased number of atoms being deposited and rightful thickness growth of nanostructured coatings of the templates. It is quite appropriate that the highest nanofilms were observed in the cases of 10% alloying additives that caused the emergence of additional 0.2–0.6 nm of layer-by-layer evaporation.

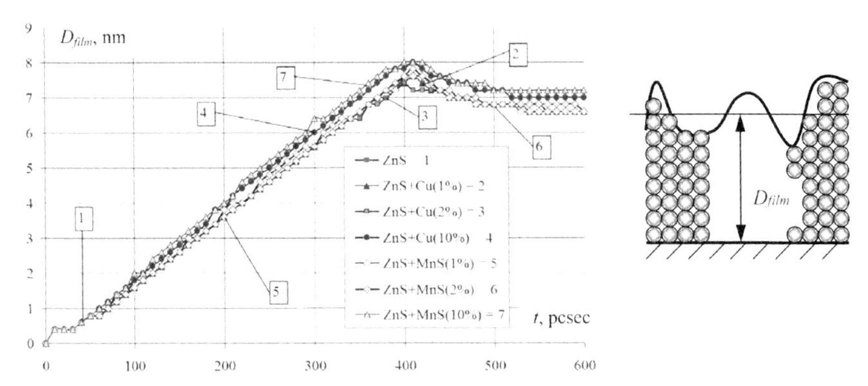

FIGURE 10.28 Average thickness of nanofilm on the template surface for different deposition angles.

10.6 CONTROL OF THE PROCESSES OF ATOM DEPOSITION AND FORMATION OF NANOFILM COATINGS

To investigate the possibility of controlling the processes of deposition and formation of nanofilms and burying of porous templates, the angle to the normal line of the porous template in which the epitaxial atoms and molecules move to the deposition surface was selected as one of the modeling control parameters. The following dimensional parameters were chosen for the given series of computational experiments: template length—19.1 nm, template width—19.1 nm, template height—11.6 nm, pore radius—5 nm, and pore depth—10 nm. The molecular ZnS in the total amount of 200,000 pieces, without the addition of any additives and alloying admixtures, was deposited onto the porous template of aluminum oxide. The template temperature was maintained at the level of normal one—293 K. The deposition was uniform over the whole template surface and with the same intensity by time. The velocity of atoms during epitaxy was 0.05 nm/ps.

The percentage of atoms in the pore relative to the total number of particles being deposited for different time moments of the condensation stage and nanofilm growth and different angles relative to the normal line of epitaxy direction is given in Figure 10.29. The graph analysis demonstrates that the deposition process only slightly depends on the angle, at which the atoms are evaporated, and it is practically identical in time. Slight deviations of the atom share in the pore are observed at the time moments of 200–400 ps, when the active rearrangement of the internal structure of nanofilms and nanoformations in the pore takes place. Afterwards, the dependencies under consideration are stabilized and reach the stationary regime. he stationary behavior of the atom percentage in the pore relative to the total amount of the particles being deposited (time moments of 450–600 ps) still has a slight difference for different deposition angles. The deviation from the normal line to the template surface results in slight increase in the area onto which the atoms are deposited that is explicitly confirmed in the graph considered. The deposition angles for all graphs are given in degrees.

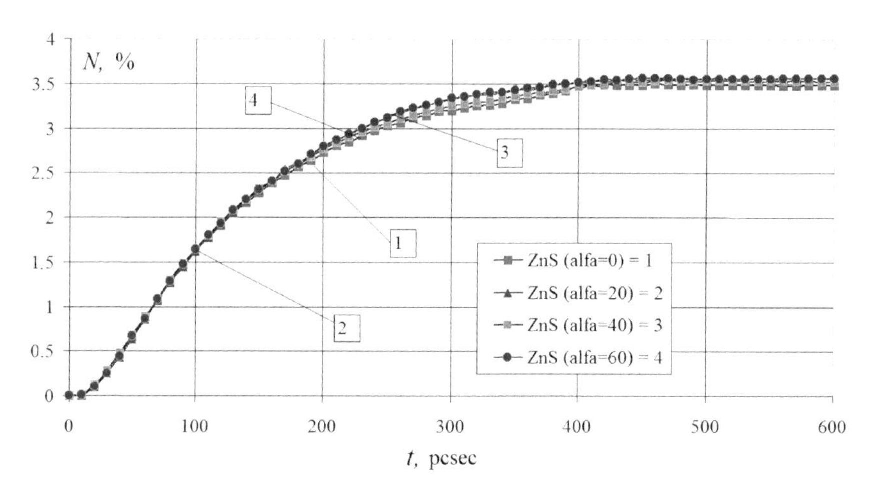

FIGURE 10.29 Percentage of atoms in the pore relative to the total number of particles being deposited for different deposition angle.

Similar behavior is observed for the position of mass center of nanostructure being formed inside the pore shown in Figure 10.30.

As we can see from the behavior of the graphs in Figure 10.30, the change in the height of mass center of the deposited atoms and molecules inside the pore also only slightly depends on the deposition angle. Slight deviations in the behavior of dependencies occur for the time moment of 30 ps of the condensation stage.

The height of nanostructures for the deposition angles of 20° and 40° is a bit lower. After the time moment of 400 ps, the situation changes and the mass center for the epitaxy angle of 0° has the least height. Thus, for the case of deposition along the normal line to the template surface, the mass center of atoms and molecules inside the pore is the lowest. Nevertheless, the deviations between the graphs in Figure 10.30 are slight and do not exceed 5%.

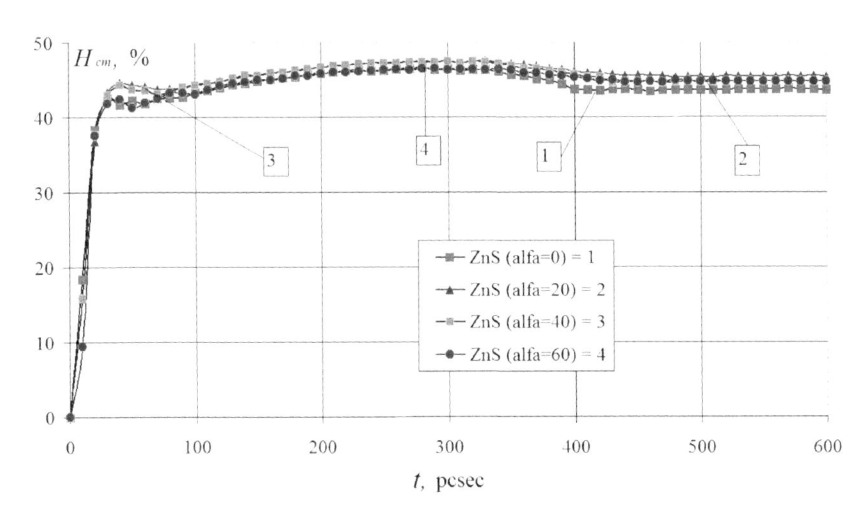

FIGURE 10.30 Relative depth of the mass center of nanoformations in the pore for different deposition angle.

To continue with the analysis, the thickness of the nanofilm being formed above the template surface was considered. The computational algorithm of the nanofilm thickness considers its layer-by-layer structure at each time moment.

The numeration of layers starts from the template surface and increases at the distance from it. The thickness of layers in the algorithm can vary, but it should depend on the crystalline structure type of the material being formed and lattice parameters, in particular, distances to the nearest neighbors. In the prevailing majority of cases, the thickness of the nanofilm intermediary layers of 0.2–0.3 nm provides the satisfactory accuracy of calculations.

For each layer, starting from the first, the number of particles in the layer and atomic density are calculated. These values are compared with similar values on the previous layer. If the spatial layers become much rarer, the computational process stops.

The nanofilm final thickness will comprise all layers previously considered. The level of atomic density is a variable algorithmic value; the value of 50% was used in this work.

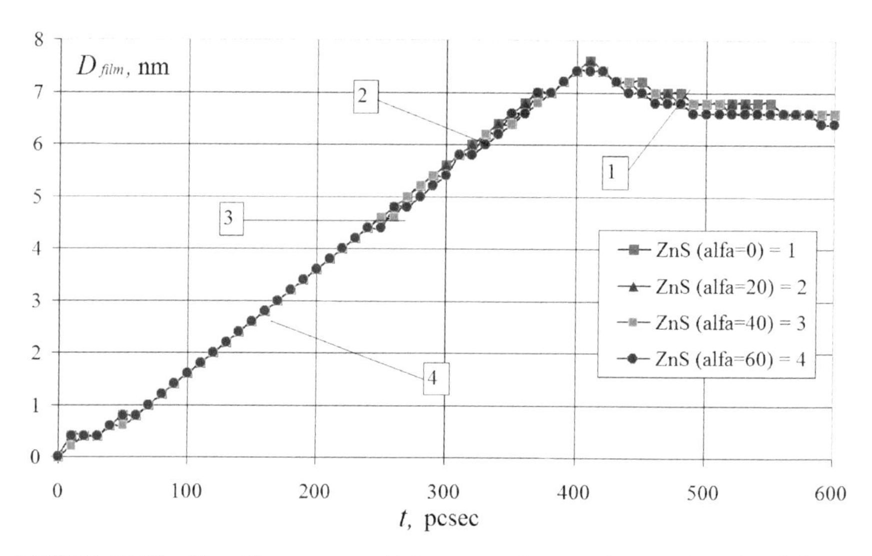

FIGURE 10.31 Nanofilm average thickness on the template surface for different deposition angles.

The nanofilm thickness growth on the template surface for different deposition angles is presented in Figure 10.31. For the initial time moments (0–400 ps), practically linear growth of the nanofilm thickness is characteristic. At the time moments of 400–600 ps, the deposited atoms and molecules are compacted and the internal structure is rearranged, so, the film average thickness slightly decreases.

The material sintering is the next possibility to control the processes of nanofilm formation and burying of porous templates. Sintering is carried out in relation to the already dusted film and is aimed at denser filling of the pore with deposited atoms. From the point of modeling, sintering is a short-duration heating of material. Because of the temperature elevation, atoms and molecules start moving more actively and the migration of particles and filling of free spaces in the internal structure can take place.

To investigate the mechanisms of nanofilm sintering on porous templates of aluminum oxide, the coatings of molecular ZnS from the previous series of computational experiments, already evaporated, were considered.

The total duration of the heating period of the substrate-nanofilm nanosystem was 200 ps. The temperature in the modeling area was maintained using a Noze–Hoover thermostat. The dimensional parameters of the alumina substrate and the pore, which cut into it, remain unchanged, as in the previous series of calculations.

To compare the results and coverage of several potential development scenarios, the nanosystem at the warm-up phase was simulated at the temperatures 293 K, 493 K, and 593 K. The temperature 293 K is normal. The nanosystem is at this temperature at the initial time of the heating stage and it is given to compare the various obtained data. As expected, at normal temperature, the system is calm, there are no significant driving forces, no significant potential differences between structural nanoelements and atoms.

Visually, atoms, nanofilms, and other composite nanostructural objects in the system behave identically for all considered temperatures (293 K, 493 K, and 593 K). There no major modifications in the nanofilms and the substrate, the pore inside the substrate is not denser filled. The obtained results are confirmed by numerical data and graphically. Figure 10.32 presents the percentage of atoms in the pore relative to the total number of deposited particles for different heating temperatures of the nanosystem.

Analysis of the graphs shows that the functional dependencies in Figure 10.32 differ insignificantly and vary slightly. In contrast to the expected results, there no the additional pore overgrowth when the temperature rises. Noticeable is the restructuring of the internal organization and structure of atoms and molecules, which are in the aperture of the substrate. At the initial moments of time, the percentage of atoms in the pore decreases, and then gradually increases. For the case of the normal temperature of the nanosystem, the restructuring of the atomic molecular structure is minimized and practically linear in time.

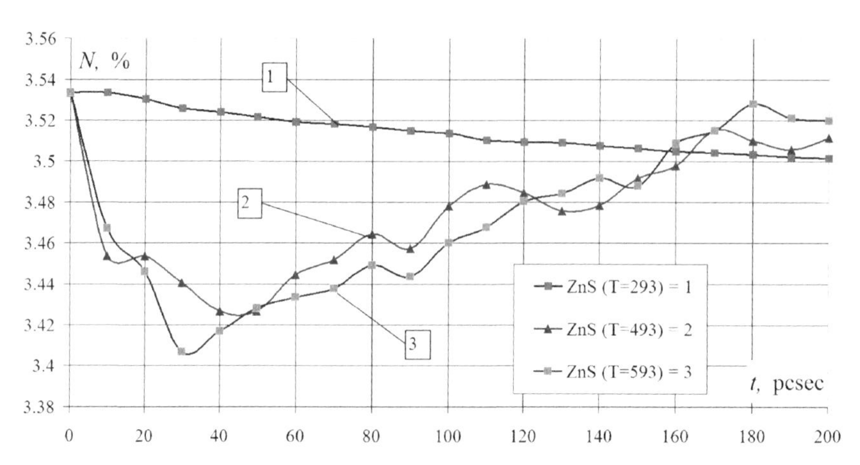

FIGURE 10.32 The percentage of atoms in the pore relative to the total number deposited particles for different heating temperatures of the nanosystem.

The relative depth of the center of mass of the nanoformations in the pore for different heating temperatures of the nanosystem is shown in Figure 10.33. The behavior of the constructed dependencies correlates with the distribution of the percentages of atoms in the pore relative to the total number of deposited particles shown in Figure 10.32.

The mass center of the nanostructures inside the pore changes insignificantly during heating. Initially, the atoms and molecules that filled the pore begin to move upward to the surface of the substrate and the formed nanofilm. At subsequent times, the position of the center of mass is corrected and tends to the initial state. It should be noted that the general dynamics and fluctuations of the quantities in Figures 10.32 and 10.33 are insignificant. The most stable behavior of the center of mass is observed for a nonheated nano system, that is, for a temperature of 293 K.

In addition to the analysis of nanoformations within the pore for the heating stage, the behavior of the formed nanofilm from ZnS was considered.

The average thickness of the nanofilms on the surface of the substrate for different heating temperatures of the nanosystem is shown in Figure 10.34.

By the results of the constructed dependences of the film thickness on the temperature, it can be asserted that heating the nanosystem to the chosen temperatures does not lead to a restructuring of the nanofilm. The thickness of the nanocoating does not increase or decrease in this case.

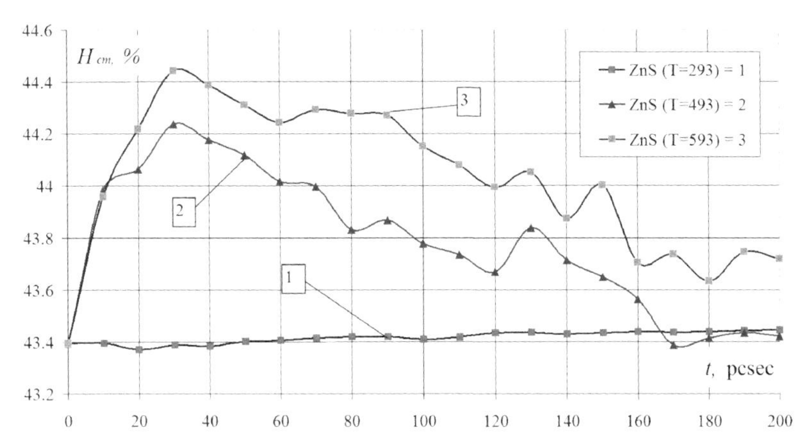

FIGURE 10.33 Relative "depth" of the center of mass of nanostructures in a pore for different heating temperatures of a nanosystem.

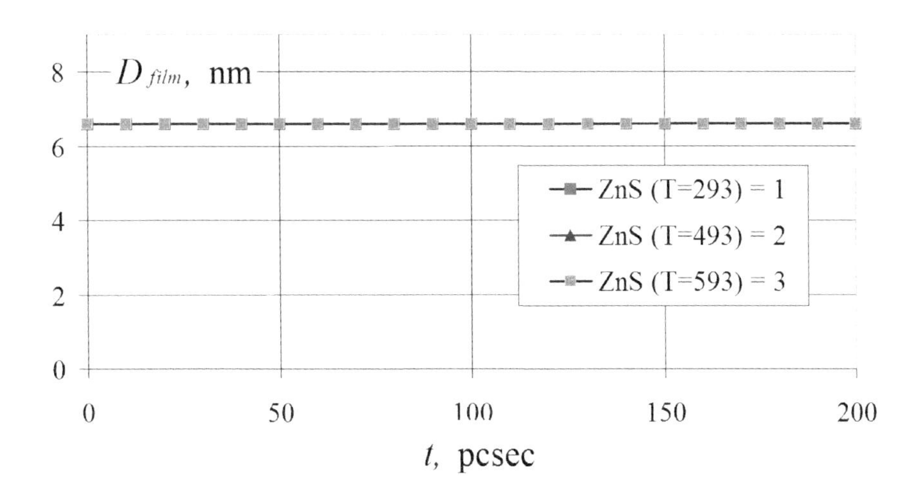

FIGURE 10.34 The average thickness of the nanofilm on the substrate surface for different heating temperatures of the nanosystem.

Investigations of the possibility of controlling the growth processes of nanofilms, overgrowing porous substrates, and the formation of nanostructured coatings have shown that the epitaxial deposition angles and heating of the nanosystem to temperatures of 293–593 K do not lead to a significant change in the formation of nanostructures. At the same time, there is a slight restructuring of the structure and internal organization of nanoelements, but there are no certain filling of the pores and changes in the properties of the nanofilms.

10.7 CONCLUSION

The setting of the problem for studying the precipitation processes of nanosized films onto the templates of porous aluminum oxide is presented. The precipitation methods of nanosized films are used for certain technological processes and applied to forecast and design nanofilm materials.

The variants of epitaxial siltation of porous templates based on aluminum oxide with various types of atoms are illustrated. Different processes of nanostructure interaction and siltation mechanisms of templates and pores are registered for different types of precipitated atoms:

(1) For the siltation with silver, gold, and magnesium atoms, the uniform pore covering with nanofilm without the penetration of theses atoms inside was observed. The nanofilm was slightly sinking in the pore region.
(2) Chromium and iron atoms demonstrated the formation of nanostructures in the air environment above the template already; the template was silted following the island principle. Small iron nanostructures were gradually growing on the template and grouping into the bigger ones. The formation of iron nanostructure was observed inside the pore.
(3) When precipitating gallium and germanium atoms, the pore was also not silted completely, the nanofilm is formed by regions on the template surface. Small gallium nanoparticles were seen on the template surface. When precipitated, germanium atoms

more intensely penetrated the pore region, but further in the process of nanofilm formation they were pulled out onto the surface.

(4) During the epitaxy of palladium, the uniform nanofilm was formed with slight sinking in the pore region. But the nonsilted opening was observed immediately above the pore during the whole condensation stage.

(5) Siltation of the template of porous aluminum oxide with titanium atoms was characterized by incoherent and coarse appearance of the nanofilm formed. The atoms got inside the pore only insignificantly and stayed on the template surface.

(6) Siltation of the template was observed for its epitaxy with vanadium atoms. A large number of vanadium atoms got inside the pore, but nonsilted regions remained on the template surface itself.

By the example of the deposition of silver atoms, it is shown that the phase transition from gas to liquid and solid state can be seen on the graph of the change in energy from temperature. The silver atoms are not separately collected into droplets and nanostructures. The phase transition occurs mixed at a temperature between the melting and boiling of silver $T = 1520$K.

Single atoms, which reached the pore bottom, were observed for all types of precipitated atoms. The most complete and dense pore siltation was registered in the process of gallium epitaxy. The pore filled with atoms can be considered as a quantum point and used to obtain optic and electric effects.

When studying the siltation with gallium atoms of the coatings with pores of different sizes, it was found out that the number of atoms actively grows in the pore in the time interval of 20–120 ps. The pore siltation after 120 ps of condensation time is accompanied by the reconstruction of atomic structure that contributes to the stabilization of dependencies and slight decrease in the percentage of gallium atoms, which penetrate the pore. Besides, the mass center of precipitated atoms is stabilized at different depths of the pore. For the pores with the radii of 2 and 3 nm, the mass center is formed above the middle of the pore depth. With the pore growth, the mass center starts to be formed in one place–near the middle of the pore depth. The further growth of the pore radius (over 5 nm) does not significantly influence the mass center; the pore is already quite compactly packed with the precipitated atoms.

The material structure (crystalline or amorphous) was analyzed based on the mean value of ideality parameter of crystalline lattice for four cases of computational experiment: solid template with the deposition of Zn and S atoms; solid template with the deposition of Zn, S atoms, and 5 % of Cu atoms; porous template with the deposition of Zn and S atoms; porous template with the deposition of Zn, S atoms, and 5 % of Cu atoms. The structure of both the template and nanofilm formed was evaluated.

In the result of investigations it was found out that ideality parameter of crystalline lattice for all four cases of computational experiment behaves similar. The parameter value both for templates and nanofilms is rather large that indicates the amorphous structure of the materials formed. The moment of nanofilm deposition completion is crucial for the structure change: if the structure rearrangement and atom movement before it was active, the change in coordinates to more energy beneficial position after it was insignificant.

During the condensation, the least value of the crystalline structure parameter was observed in the computational experiment without a pore with the addition of 5% of copper atoms. The stabilized evaporated material had the least parameter value for the template without a pore and without the addition of copper atoms.

The modeling results can be used when developing and optimizing technological processes of optic coating formation and analysis of physical properties of nanofilms and nanostructures formed in pores and on surfaces of aluminum oxide templates.

KEYWORDS

- **epitaxial deposition**
- **mathematical modeling**
- **modified embedded atom method**
- **nanofilms**
- **nanostructures**
- **porous templates**

REFERENCES

1. Kelchner, C. L.; Plimpton, S. J.; Hamilton J. C. Dislocation Nucleation and Defect Structure During Surface Indentation. Phys. Rev. B. **1998,** *58* (17), 11085–11088.
2. Plimpton, S. Fast Parallel Algorithms for Short-Range Molecular Dynamics. J. Comp. Phys. **1995,** *117*, 1–19.

Index

A

Abell–Tersoff Potential, 108–109
Absorption spectral analysis, 73
  and optical band gap, 191–197
    full width at half maximums (FWHM), 195
    optical thicknesses of ZnS:Cu (Mn) films, 192
Adams numerical methods, 119–121
  deposition problem
    numerical methods for solving, 119–121
Aluminum oxide films
  processes running, 21
  types of, 19–21
Anodic aluminum oxide (AAO)
  influence of anodization parameters on structure
    distance between pore centers, 25–26
    pore diameter and porosity, 26–27
    thickness, 28–29
  semiconductors nanostructures, 29–33
    MOCVD, 30
    SEM patterns of ZnS nanotubes, 30
    TEM and SEM patterns, 32
    thermal deposition of semiconductors, 32, 33
  structure, 22–23
  synthesis of, 141–151
    anodizing cell, 143
    chronoamperograms, 149
    coulometric control thickness, 143–145
    dependence of the average pore diameter, 147
    microphotograph, 148, 150–151
    pore diameter control, 146–151
    SEM images, 147
Anodization parameters on structure
  influence of
    distance between pore centers, 25–26
    pore diameter and porosity, 26–27
    thickness, 28–29

B

Born–Karman periodic boundary, 135
Buckingham potential, 107

C

Chemical bond characters
  chemical composition, 165–175
    Cu-doped samples, XPS, 166–171
    Mn-doped samples, XPS, 172–175
    2p spectra, 168–169
    Zn, S, and Cu concentrations, 170

D

Deposition of doped ZnS
  Cu and Mn onto smooth templates, 159–163
    and amount, 161–163
    list, 161–163
    nomenclature, 161–163
  ultrahigh-vacuum evaporation, facility, 152–159
    analytical chamber USU-4, 153–154
    chamber of ultrahigh-vacuum, 154–155
    interchamber accessories, 157
    Knudsen evaporation cell, indirect-heating, 155
    software controlling evaporation processes window, 156
    stand with control equipment, 158
    template holders, 157
    template-loading chamber, 152–153
Deposition problem
  numerical methods for solving, 113–125
    Adams numerical methods, 119–121
    Euler method, 114–117

predictor-corrector methods, 121–123
Runge-Kutta methods, 117–119
Verlet algorithms, 122–125

E

Electroluminescence, 197–202
emitters, illuminating characteristics, 2, 78–80
dependence of main parameters of ELS, 80
heterojunction, 79
spectral, 78
mechanism in ZnS ELS, 5–6
excitation of electroluminescence, 5
freed electrons, 6
phenomenon, 2
photo images, 198
ZnS:Cu samples
phosphor structure energy layers, 200
spectra of, 199
ZnS:Mn phosphor structure energy layers, 202
Electroluminescent light sources (ELS), 3–5
devices and methods, 81–83
diagram of TFELS, 4
thin-film electroluminescent screen, 4
Electrophosphors, 51
Embedded atom method (EAM), 109–110
Emitters, illuminating characteristics
electroluminescence, 2, 78–80
dependence of main parameters of ELS, 80
heterojunction, 79
spectral, 78
Euler method, 114–117
deposition problem
numerical methods for solving, 114–117
Evaporation and deposition processes, 37–43
crystal surface, 38
experimental approaches and equipment
evaporator, 45–48
setup for vacuum-thermal deposition, 43–45
Frank–van der Merve growth, 39
kinetic theory, 38
OR stage, 41
Extended X-ray absorption fine structure (EXAFS), 55–61
formation of the diverging electron wave, 57
local atomic structure, 181–188
Cu-doped samples results, 181–184
Mn-doped samples results, 184–187
XANES investigations results, 187–188
model-dependent approach, 60
path fitting, 59
physical basics, 56–57
signal, 58
stages of processing experimental, 59

F

Frank–van der Merve growth, 39
Full width at half maximums (FWHM), 195

I

Influence of anodization parameters on structure
distance between pore centers, 25–26
pore diameter and porosity, 26–27
thickness, 28–29
Intermolecular potential
angle between three atoms, 99
dihedral angle between four atoms, 100
distance between two atoms, 98

K

Knudsen evaporation cell
indirect-heating, 155

L

Lennard-Jones potential, 102–103
Light-emitting electroluminescent panels, 1
Local atomic structure
EXAFS spectroscopy data based, 181–188
Cu-doped samples results, 181–184
Mn-doped samples results, 184–187
XANES investigations results, 187–188
Luminescence, 2

M

Main peak wavelengths, 75
Metal–organic chemical vapor deposition (MOCVD), 30
Mi potential, 106
Modified embedded-atom method (MEAM), 111–113
Monochromator
  diffraction grids, 73–76
    main peak wavelengths, 75
    measurements of transmission spectra, 76
    optical scheme, 74
    white light dispersion, 75
Monofilms
  porous alumina, formation, 208–220
    depth of mass center of atoms, 220
    kinetic energy dependence, 209
    percentage of precipitated atoms, 219
    precipitated elements, physical properties, 210–212
    siltation of the template, 214–218
    silver atom precipitation results, 213
    temperature dependence, 209
Morse potential, 105–106

N

Nanofilm
  deposition, 205–208
    pore size into substrates on nature effect, 220–223
    problem solving steps, 206
    structure of templates, investigation, 223–228
    system being modeled, 207
  formation
    ab initio method, 86–89
    atom deposition processes, control, 235–242
    calculated volume increase, 95
    integration step, 96
    mesic media, nanostructure movement in, 92–97
    modeling methods of, 85–97
    molecular dynamics (MD), 89–91
    multicomponent, 229–235
    mutual arrangement of, 93–94
    porous aluminum oxide, admixtures on templates, 229–235
    precipitation processes, 242
    and semiempirical methods, 86–89
    types of precipitated atoms, 242–243
Nanostructures morphology
  SEM based data, 176–177
Numerical methods for solving
  deposition problem
    predictor–corrector methods, 121–122

O

Optical band gap
  absorption spectral analysis, 191–197
    full width at half maximums (FWHM), 195
    optical thicknesses of ZnS:Cu (Mn) films, 192
Optical transmission spectra
  determination of band gap of semiconductors, 76–77
Oswald ripening (OR), 41

P

Paper, 9
Periodic boundary conditions, 133–136
  use, 136
Porous structure formation
  mechanism, 23–25
Potentials of force fields, 97–113
  Abell–Tersoff Potential, 108–109
  Buckingham potential, 107
  EAM, 109–110
  intermolecular potential, 97–102
  Lennard-Jones potential, 102–103
  MEAM, 111–113
  Mi potential, 106
  Morse potential, 105–106
  Stillinger–Weber potential, 107–108
  Stockmayer potential, 104–105
Predictor–corrector methods
  deposition problem
    numerical methods for solving, 121–122

R

Runge-Kutta methods, 117–119
deposition problem
numerical methods for solving, 117–119

S

Semiconductors nanostructures
anodic aluminum oxide (AAO), 29–33
MOCVD, 30
SEM patterns of ZnS nanotubes, 30
TEM and SEM patterns, 32
thermal deposition of semiconductors, 32, 33
Stillinger–Weber potential, 107–108
Stockmayer potential, 104–105
Synthesis of anodic aluminum oxide (AAO), 141–151
anodizing cell, 143
average pore diameter
dependence of, 147
chronoamperograms, 149
coulometric control thickness, 143–145
microphotograph, 148, 150–151
pore diameter control, 146–151
SEM images, 147

T

Thermodynamic parameters
and energy of nanosystem, 125–133
Berendsen algorithm, 128–129
Berendsen barostat, 132
collision thermostat, 127–128
energy parameters, 125–127
Friction thermostat, 130–132
general thermodynamic, 125–127
Nose–Hoover thermostat, 129–130
Parrinello–Rahman barostat, 132–133
Thin-film electroluminescent sources (TFELS), 3–5
application of dielectric layers, 4–5
diagram, 4
layout diagram, 4
structures, 3
transparent metal electrodes, 3

U

Ultrahigh-vacuum evaporation
facility, deposition of doped ZnS, 152–159
analytical chamber USU-4, 153–154
chamber of ultrahigh-vacuum, 154–155
interchamber accessories, 157
Knudsen evaporation cell, indirect-heating, 155
software controlling evaporation processes window, 156
stand with control equipment, 158
template holders, 157
template-loading chamber, 152–153

V

Verlet algorithms, 122–125
deposition problem
predictor–corrector methods, 122–125

W

White light dispersion, 75

X

X-ray absorption near-edge structure (XANES), 62–64
investigations results, 187–188
X-ray diffraction (XRD)
investigations, 51–55, 178–181
EXAFS, 55–56
influence of defects, 53
magnified image, 180
Mn-doped samples, 180
operation scheme, 52
patterns of Cu-doped samples, 178
X-ray photoelectron spectroscopy (XPS), 64–68

Z

ZnS:Cu samples
electroluminescence
phosphor structure energy layers, 200
spectra of, 199
ZnS nanostructures
ELS

obtained by printing emission layer, 7
structure, 8
with working layers, 12
phosphors based on doped, 6–15
"bulk" samples, 12
colloid particles, 8
electroluminescence, 9
energy levels in ZnS, 15
graph of dependence, 11
$Mn2^{+}$-doped ZnS nanoparticles, 13
molar ratio, 8
number of alternating layers, 9
paper, 9
solution, 10
X-ray powder diffraction studies, 13